Architecting Enterprise AI Applications

A Guide to Designing Reliable, Scalable, and Secure Enterprise-Grade AI Solutions

Anton Cagle
Ahmed Ceifelnasr Ahmed

Apress®

Architecting Enterprise AI Applications: A Guide to Designing Reliable, Scalable, and Secure Enterprise-Grade AI Solutions

Anton Cagle
Sun Prairie, WI, USA

Ahmed Ceifelnasr Ahmed
Myrtle Beach, CA, USA

ISBN-13 (pbk): 979-8-8688-0901-9
https://doi.org/10.1007/979-8-8688-0902-6

ISBN-13 (electronic): 979-8-8688-0902-6

Managing Director, Apress Media LLC: Welmoed Spahr
Acquisitions Editor: Celestin Suresh John
Development Editor: Laura Berendson
Coordinating Editor: Kripa Joseph

Cover designed by eStudioCalamar

Cover image designed by Freepik (www.freepik.com)

Distributed to the book trade worldwide by Springer Science+Business Media New York, 233 Spring Street, 6th Floor, New York, NY 10013. Phone 1-800-SPRINGER, fax (201) 348-4505, e-mail orders-ny@springer-sbm.com, or visit www.springeronline.com. Apress Media, LLC is a California LLC and the sole member (owner) is Springer Science + Business Media Finance Inc (SSBM Finance Inc). SSBM Finance Inc is a **Delaware** corporation.

For information on translations, please e-mail booktranslations@springernature.com; for reprint, paperback, or audio rights, please e-mail bookpermissions@springernature.com.

Apress titles may be purchased in bulk for academic, corporate, or promotional use. eBook versions and licenses are also available for most titles. For more information, reference our Print and eBook Bulk Sales web page at http://www.apress.com/bulk-sales.

If disposing of this product, please recycle the paper

To Nicole, my soulmate.

Anton

To my extraordinary parents, whose lives are a testament to resilience, sacrifice, and unconditional love.

Dad, who at the age of 13 made the selfless decision to leave school and take on the responsibility of providing for his family. Your courage and dedication have been the cornerstone of our family's strength. Mom, though you never had the opportunity to attend school, your wisdom, kindness, and tireless dedication to raising us have shaped the core of who we are. You both poured your hearts into raising nine children, instilling in us the value of education and the importance of perseverance. Because of you, we each proudly hold college degrees, a testament to the dreams you nurtured within us.

To my big family and friends in Egypt, and my family in the United States – the Swangers, the Paszkowskis, the Gipes, and the newly married McCloskeys – thank you for being my steadfast support system, my source of inspiration, and my constant encouragement throughout this journey.

And to Melinda Dean, thank you for your encouraging presence in my life. I am grateful for the moments we've shared.

With all my love and deepest gratitude,

Ahmed Ceifelnasr Ahmed

Table of Contents

About the Authors

Anton Cagle is a seasoned leader specializing in cloud automation and AI Ops, boasting over two decades of expertise in enterprise architecture and application design. With a passion for delivering democratized, data-driven solutions and automation, Anton focuses on empowering medium- to large-sized companies. His dedication extends to mentoring and coaching engineers at all skill levels, fostering a culture of continuous learning and innovation.

Recognizing the pivotal role of cloud, data, and AI in shaping the future of business software, Anton is on a mission to guide companies beyond basic automation solutions. His goal is to seamlessly integrate big data and machine learning into organizational frameworks, preparing businesses for the next wave of scalable operations. Anton's approach has led to remarkable transformations for clients, including the reduction of deployment process waste, accelerated feature time to market, and the implementation of cutting-edge cloud data architectures.

Ahmed Ceifelnasr Ahmed is a highly skilled ML engineer, data scientist, and cloud engineer with over six years of experience in developing and deploying data-driven solutions. Ahmed specializes in building and fine-tuning machine learning models, leveraging advanced deep learning techniques, and optimizing cloud-based solutions. His expertise extends to cloud engineering and DevOps practices, where he excels in designing and implementing scalable, efficient cloud architectures and automating deployment processes.

With hands-on experience in AWS cloud environments and a strong background in cloud tools, Ahmed is adept at integrating artificial intelligence with cloud technologies to create robust, production-ready solutions. He has a proven track record of driving impactful results across various industries, from retail and real estate to fitness and enterprise applications.

Ahmed is committed to continuous learning and growth, always seeking to make a significant impact in the fields of AI, data science, and cloud engineering. His career reflects his dedication to advancing technology, optimizing cloud infrastructure, and fostering innovation through data-driven strategies and cutting-edge technology.

About the Technical Reviewer

Dave Grundgeiger is an author, speaker, and the founder of Contextium, a company dedicated to making AI more accurate and usable for a wider range of applications, in part, by detecting AI hallucinations. He has 35 years of experience as a software engineer.

Introduction

If you are interested enough in AI to consider reading this book, then we are sure you are familiar with the massive discrepancy in the different kinds of material on AI that has been released in the past few years. On one hand, you have academics who are concerned with the mathematics behind the creation of new AI models. These come in the form of white papers with a prerequisite in reading mathematical symbolism to understand and thick books on machine learning that show you how to implement your own AI models through the use of specialized programming libraries.

On the other end of the spectrum, there are a million marketers with YouTube channels and "lead generating" ebooks with titles like *Ten Tricks to Use AI Prompting to Reach Your Goals (Plus a Bonus Tip If You Watch to the End!)*. AI really has made a huge impact in the world of content marketing, but these videos and ebooks don't help engineers and business people who are tasked with building large systems that meet all of the demands associated with mission critical applications.

The aim of this book is to help businesses of all sizes build AI applications that are scalable, reliable, and sustainable. We approach this book with an engineering perspective, and our audience will have a solid engineering mindset, but AI brings with it an unprecedented leap in capabilities in the automation of business process that used to require entire IT departments to build and support. Businesses no longer need a large corporate IT team to create and manage their enterprise applications – many business processes can now be automated and systematized with a handful of employees with a strong grasp on business process and an engineering mindset.

With that said, this book is written for anyone who is concerned with designing large-scale Enterprise AI. That does not mean that you need to have a large team of programmers, it means that the systems our audience is designing are of sufficient complexity to justify a significant consideration to the architecture of the system. Small, single-purpose applications should not spend large efforts on their architectures. Multifaceted enterprise systems, however, ignore their architecture to their detriment.

One of the most common issues facing architects, business leaders, and engineering leaders who are faced with building AI systems is understanding the problem domain, tools, and emerging patterns enough to feel confident in what they are building. This book is meant to provide that overview of the AI problem space enough to be a solid launching pad for building successful AI systems.

This book is not a programming book. As such, we do not dive into specific frameworks such as scikit-learn, Keras, or specific machine learning algorithms. We do touch on tools and frameworks in Part 2 but only as a means for further exploration.

This is also not a prompt engineering book. As such, you will not find prompt recipes, but we will explain how generative AI prediction models conceptually work, which can be used as a mental framework for writing suitable prompts and have a deeper understanding of the reasons for why the Generative Pretrained Transformer (GPT) models respond the way they do.

We believe that a wealth of technical resources about programming tools and prompt engineering are available all around us, much of it for free. In fact, there is so much of it that we have a hard time digging through it and keeping up with the latest advancements while throwing away the rapidly aging frameworks that are no longer relevant. This book is unique in that it covers the conceptual and practical approaches to Enterprise AI that serve business leaders from an engineering perspective. This is not a book written from an executive business school perspective but rather from the perspective of hands-on, seasoned professionals like you.

How This Book Is Organized

This book is split into four parts, each of which could be read independently. The first and the last part act as conceptual bookends that hold together the two more technical parts in the middle. As such, those who are more interested in the business problems that can be solved by AI, along with their social and ethical side effects, can skip the middle chapters and read Parts I and IV.

Technical leads and developers who have been given a problem to solve and do not have much influence over the organizational and legal implications of their applications may comfortably stick to the middle technical chapters that focus on building and maintaining AI applications.

Part 1: Defining Your AI Application

One of the biggest challenges business teams are faced with is finding the right fit for AI within their enterprise. In many cases, the technology solutions for AI exist for a number of business purposes, but business leaders struggle to envision how they could practically incorporate them into their technology solutions. The first part provides a conceptual framework that helps business and engineering leaders understand the key benefits and limitations of modern AI. Understanding what AI is good at, how it comes up with its predictions, and how humans can play an essential role in the AI-powered enterprise are key to envisioning new AI solutions that create a competitive advantage in the marketplace and not just satisfy executive leadership's directive to just "start using AI."

Part 2: Designing Your AI Application

In this part, we introduce an enterprise framework for building multi-agent AI applications. We also provide an overview of the technical foundations of machine learning and large language models. This foundation will provide us with the building blocks we need to understand AI agents.

Part 3: Maintaining Your AI Application

This part discusses the critical issues of data curation, test automation, and security that are required to maintain AI applications that will successfully grow and evolve over time, instead of being shelved or torn down at the first sign of a service disruption or required upgrade.

Part 4: AI-Enabled Teams

In our final part, we discuss the organizational impacts of AI applications, from AI agent team members to legal policies. The organizational impact of AI on business and engineering teams has the potential to be incredibly disruptive. An understanding of this potential disruption before we move to implement our AI applications can help us address the social issues at hand before they become a serious problem in production, or perhaps never even make it to production because we did not get the proper buy-in and support from our critical business partners.

AI Architecture As Evolution

We approach this book with the understanding that the technical tools we need to build Enterprise AI applications are available today, and the technical hurdles to creating them, such as the need to have a team of expert programmers, are rapidly decreasing. We see agent-based AI architectures as an evolution of existing enterprise architectural patterns that have been around for more than a decade. In this book, we do not look at AI as a dream of technology that only the elite can afford to build. One of the biggest reasons why there is so much hype around AI is because it is much cheaper to implement and much more powerful (when focused on the right problems) than most people assume. With that in mind, we hope this book will serve you well and provide a solid platform for conceptualizing and building your Enterprise AI applications.

Anton Cagle
Ahmed Ceifelnasr Ahmed
Fall 2024

PART I

Defining Your AI Application

CHAPTER 1

Human Flexibility and AI Specialization

Overview

This chapter begins our attempt to lay a conceptual framework for designing Enterprise AI applications. We begin by discussing the high-level limitations and costs of implementing AI in an enterprise and then continue the discussion through examples of human ingenuity to understand how we can focus our use of AI to design useful Enterprise AI applications. In the second and third parts of this book, we will lay out a technical framework for building and maintaining AI applications.

Janitors and Engineers

"Janitors will be here long after the scientists have been replaced by AI."

A dear friend of mine, Matt, who has run a number of Chicago businesses from restaurants to renovation services has his doubts about AI.

"We were promised Rosie the Robot. Instead, the engineers who worked on that problem seem to be working themselves out of a job."

© Anton Cagle, Ahmed Mohamed Ceifelnasr Ahmed 2024
A. Cagle and A. M. C. Ahmed, *Architecting Enterprise AI Applications*,
https://doi.org/10.1007/979-8-8688-0902-6_1

This is a bit of an overstatement, and no doubt he was excited to rib his engineering friend, me, about his choice of a career over mine. He makes some good points though, and most of our thinking about so-called "knowledge work" is about to be refined.

Let's follow Matt's line of reasoning.

Consider a fictional handyman named Harry. Harry the Handyman and Rosie the Robot are both called to fix a leaky toilet. Upon arriving on the scene, our two protagonists, Harry and Rosie, are presented with water pooling all around the toilet and bathtub, clearly coming from a constant trickle of water running down the back of the toilet. Upon further inspection, it seems that the homeowner has tried to perform his own fix on a leaky pipe by removing a large piece of piping from under the commode with his own handiwork. The makeshift solution consists of a piece of pipe that has been attached with a fitting that is a quarter inch too big, then wrapped with a piece of black foam insulation to help with the condensation from the cold pipe.

For Harry, this is an easy fix–pull off the insulation, remove the ill-fitting piece, and replace it with one of the right size. For Rosy, this task is much more difficult. First, she must recognize that the problem pipe is actually hidden under a large piece of black insulating foam that masks the underlying problem. The work was done by the homeowner and does not follow standard convention, so there will need to be some improvisation on the part of Rosy. In fact, it would just be easier for Rosy to just remove the entire toilet and replace it with a standard model with all standardized pieces, including the leaky pipe. This is a much more invasive and expensive solution than just replacing a single pipe. For Harry, this job would take about 30 minutes to fix with little fuss. For Rosy, however, the recommended solution of a full toilet replacement would likely take a couple hours and a lot of carting of wet toilet pieces through the house. Even at Harry's premium plumber's rate, it is much cheaper and less messy to hire him than to bring in Rosy.

Our lighthearted example above uses a real-world scenario, plumbing, to make the point easy to grasp. It turns out that most aspects of the skilled trades (home services, construction, landscaping, etc.) rely on human flexibility and ingenuity. Almost all service work done inside the home requires the service professional to deal with some level of customization. Work that is done outside, such as tree trimming or new home construction, regularly must deal with the dynamic factors of terrain and weather. For this reason, most robotics outside of a sterile lab environment today are highly supervised and essentially work as an augmented pair of hands and don't really qualify as true artificial intelligence as we use the term today, though there are pieces of machine learning techniques present in many of these tools.

This book deals with AI in the digital space and more specifically in the corporate enterprise environment where the tools work in support of or in tandem with enterprise knowledge workers. Enterprise solutions that may utilize AI involve the use of information in a specific business context in order to perform actions or make decisions that work toward reaching specific corporate goals. These goals are tied to a business vision that is usually always tied, but not limited, to a financial incentive. Other incentives could include moral outcomes, political outcomes, or even just the increased social status of the business or the owner himself.

AI is particularly good at consuming information and then performing some work based on that information. For those who haven't spent a few years in the trenches of the chaotic world of Enterprise IT, and who have watched a few wiz bang demos of the new "AI Developers" available on the market (I'm looking at you, Devin!), it seems like the role of the application developer is about to be replaced. I would like to suggest, however, that the problems that AI faces in corporate IT have much more in common with that of the handyman than some of our Silicon Valley Sales Bros would have us think.

Let's take another fictional example similar to our plumbing example above but now recontextualized into a corporate IT environment. Our protagonists this time are Albert the AI and Debbie the Developer.

XYZ company has a problem integrating its CRM system to its website. The website has recently been overhauled to account for a new rebranding effort by the marketing department, which includes a new set of integrations that works with a new retail partner. These new changes seem to have broken the connections between the commercial CRM system and the company's custom website. Bringing Albert the AI in to fix the integration is the first thought that the managers have, of course, because they don't have to worry about hiring a developer to do this work. The problem that the AI has, first, is understanding the problem–what needs to be fixed? If we allow for some good use cases on the part of the company leadership staff, the next question is finding the code that integrates the two systems. Since the website has a number of custom backend components to tie it to all of its systems, Albert will need to somehow identify each of the code modules used in integration and then write code that can be tested and checked for validity. To make it trustworthy, the code needs to be readable and validated by a human. The issue, of course, is that these custom integrations are written by individual humans, perhaps several humans over a long period of time, and contain multiple software versions and dependencies that need to be understood and tested by Albert the AI. This is all possible, but very expensive, and the likely outcome would be to just replace any custom components outright.

The more straightforward solution would be to hire Debbie the Developer to fix these integrations for you. Just like the pipe under the insulation in our example above, Debbie's role as a developer is to peel back the layers of custom code to discover the root of the broken integration. She may even suggest pieces of code that could be completely refactored or gotten rid of. The timeline for this work is usually orders of magnitude cheaper than replacing whole systems. Over time, large companies have hundreds of these custom integrations, and while I am not suggesting that a move to standardized platforms is never in a company's best interest, a company will always have value-added customizations to its software that will be differentiators. In fact, even commercial off-the-shelf software avoids complete standardization of its products for business reasons.

Software companies call these custom differentiators their "moat." These differentiators stop other companies from copying their IP, and often this comes at the integration layer for core pieces of functionality. More traditional companies often refer to their core differentiating functions as their "flywheels" in reference to Jim Collins' classic book *Good To Great*. For insurance companies, for instance, their flywheel system may be their actuarial algorithms and custom analytical tools. For a food delivery service, their combined moat and flywheel is likely their custom mobile app used by delivery drivers. While it could be copied by others, it is just complex enough to make it significantly inconvenient for competitors to do so. In addition, there are likely several pieces of unique intellectual property in place that prohibit competitors from recreating large pieces of app functionality.

Building AI applications that can work within the dynamism and complexities of unique Enterprise IT environments can be an expensive endeavor if done without forethought and planning. Understanding both the possibilities and the limitations of AI can not only help you avoid the embarrassment of a failed project but also help you identify appropriate opportunities for Enterprise AI applications, for which there are plenty. The field of corporate IT today is ripe for the kind of business disruption that AI may bring. Small businesses are now able to perform activities and analyses within their corporate operations that only Fortune 500 companies could afford to do just a few years ago. Independent developers armed with AI tools are able to increase their productivity and reach several times over. The goal of this book is to help developers and business leaders think about Enterprise AI applications in a way that is thoughtful, forward thinking, and above all, practical.

Defining Purpose vs. Information Summarization

The recent talk about data-driven decision-making has made many people believe that AI will naturally take the place of corporate decision makers. The logical jump seems obvious–AI utilizes data to make decisions much faster and accurately than humans. There is a fundamental missing step here that most well-meaning data analysts (among others) miss. Computers are good at following directions, and AI systems are getting better at following general directions, but the business directives themselves at the highest levels are complicated and multifaceted.

Businesses that operate with the expectation of making a profit have a high-level goal of making money. Each business focuses on a very specific set of products or services. So, making money selling sweatpants is a very good goal. Business strategy consists of more than just this, though. How the company differentiates itself among 20 other sellers of sweatpants is part of how the business plans to make money. Is it the low price or high fashion leader in the marketplace? Does it actively market to a specific niche or does it service a broad audience?

One could say that being data-driven means that algorithms could do a much better job identifying a target market, price range, and clothing line than humans could. First, of course, this assumes that there is an existing market from which that data can be gleaned for the AI to do its work, but we will set that consideration aside for this example. It is likely that an AI can make good recommendations toward how to make the most money selling sweatpants that are just slightly more profitable than the competition. However, before we set this AI loose to implement our new product and operations strategy, we should consider some fundamental factors.

When we think of business directives at this level, we usually think of business vision. Typically, this vision includes phrases such as "delight our customers" and "provide world class service." These statements are intentionally vague, and a good company vision will provide enough direction to communicate its focus while also being broad enough to encompass novel approaches in implementation. In reality, there are a number of unspoken factors in these vision statements that are easily understood by humans but may easily be missed by AI. For instance, an AI could discover that based on previous data it would be very profitable to focus a product on very specific demographic. If that demographic focus discriminates against other protected demographics, this is a morally unacceptable strategy.

Complex human morality is an unspoken undercurrent that defines much of our business practice. Proponents of the efficient market theory of economics may argue that the competitive prices of the market itself reflect the general moral sentiment, but in my experience good business leaders do take ethical behavior seriously, especially if it is unspoken.

In the early 2000s, I worked for a large life insurance company out of the midwest. We offered an accidental death and dismemberment program that was attached to a credit card. The product was offered at the issuance of the credit card with the guarantee that as long as the card did not hold a balance, there would be no fee for this insurance product. To the consumer opening the credit card, this had all the makings of a free product. My boss at the time, Sammy, was our chief actuary. When our division was expanded, he took over this product among others. It turns out that this was an extremely profitable product for the company, pulling in millions of dollars per year with very little overhead. It was basically "passive income" for us.

Sammy took his personal morality very seriously and was an active member of a local church. He took particular interest in this product, and over the period of a few months asked me for a lot of analytical data on it. Supporting this product had become a significant part of my daily activities during that time, so I was shocked when he came to my desk one day and announced that we were shutting it down. He had gone to our board of directors and showed them that not only do we pull in millions of dollars in premium every year (after all, most Americans actually carry a balance on their credit card), but the insurance fine print was so specific that very few customers could actually make a successful claim against it. One could make the claim that selling this kind of product is bad business because it would eventually be discovered by the press and be very bad PR for the company, lowering its value in the marketplace. Statistically speaking, these kinds of products are very rarely discovered, and it would be hard to believe that an AI that has been optimized to "generate profits within our customer base without breaking the law" would really pass the test for most business leaders' definition of good business practice.

Profit isn't the only incentive for many companies, even if they aren't nonprofits. Several years ago, I worked as a software architect for a health insurance company that was owned by a Catholic hospital system. The Catholic hospital was a nonprofit, but the health insurance division was classified as a for-profit division of the company. One of our jobs during this time was to manage the creation of a number of new benefit plans that were introduced as part of the Affordable Care Act (ACA). Many of these plans were much less profitable for the company as we needed to cover procedures and customers who did not fall under our actuarial guidelines for profitability. For instance, one of the major reforms that was enacted in the ACA was the requirement for insurers to cover pre-existing conditions. From a pure profit perspective, this had a significant impact on our bottom line, as it did other insurers. We tried to make up for this loss with more profitable and expensive private plans, but ultimately we lost profitability on this. If profit was our only motive, this would have been very difficult to take as a business. As it was, we were part of a larger health provider network, primarily made up of hospitals and clinics, and our number one goal as a larger health provider network was to provide excellent health care to our customers. While the for-profit insurance division did suffer some short-term profit loss, the goal of the larger healthcare system was to provide for the health of the communities it served.

If the AI was implemented to make decisions based on profitability, it would have identified plans that were over-optimized toward this local goal and misaligned with the larger system goal of providing health care to customers within the larger healthcare system. There were a number of benefit packages and operation improvements that were turned down at the time as our profitability goals were throttled to stay within a certain range. If the health insurance arm of the health system became too profitable, there was an ongoing concern from our board of directors and investors that the insurance division was focusing strongly on profit over people, so there were a number of initiatives that would have been very appropriate to implement in insurance companies in the property and casualty sector that were not acceptable in the realm of healthcare insurance.

In a similar fashion, long-term versus short-term strategies for profitability are a constant balance for business owners. The life of a company must be balanced over a longer time period, and the incentive structures are constantly shifting due to opportunity costs in the market, investor sentiment, cash flow, and future product vision. One of the primary criticisms of public companies on the American stock market is that they tend to prioritize short-term profits over long-term stability due to constant investor demand for strong quarterly earnings.

A good leader who can overcome this short-term investment focus will be able to communicate a strong vision for the company that goes beyond immediate financial gain and allows investors to visualize a future company that is stronger and better than what it is today. This kind of communication requires a good deal of creativity and a deep understanding of complex human sentiment on the part of company leadership.

Imagining Ideal States vs. Interpreting Existing Information

A data-driven AI strategy can be optimized to find opportunities for profitability based on the information that currently exists in the market and in private operations data. In this way, AI can be particularly good at recognizing and extrapolating opportunities for growth based on what is already in the marketplace. There may be trends, for instance, of a particular kind of product specialization and demographic sector, that match up in unexpected ways. These opportunities can be very valuable to business decision makers, and new areas of product testing could be opened up and explored based on these insights.

These areas in which AI-driven decision-making is very competent fall into the area of pattern recognition. Utilizing AI applications for highly complicated areas of specialization is where you will find the most value. Optimizing in an area of work that has a clear definition of optimal outcome will bring major improvements to business operations.

The innovations in AI optimization will come in a similar fashion as the optimizations made when introducing lean, six sigma, and agile techniques into your business processes and project teams. Toyota paved the way for innovations in this area with its Kaizen principles in the 1980s. Toyota manufacturing plants were restructured to empower individual line workers to interrupt the manufacturing line any time an inefficiency or error was spotted in the process. The ability to empower line workers was a significant part of Toyota's success through improved quality in the 1980s and 1990s. Business leaders in the 1990s began implementing many of these same principles through programs such as Total Quality Management (TQM). Software developers in the early 2000s utilized these principles to create the agile movement, which is based on small democratic teams of software developers.

Each of these process improvement initiatives assumes an existing process and a clearly defined business vision and direction in order to identify inefficiencies, defects, and miscalculations and move within a forever state of continuous improvement. In Part 2 of this book, we will be showing you how to create software that works within the framework of continuous improvement optimization, which we believe will bring the most significant and long-lasting results from AI technology.

Many of my agilist friends will take issue with my statement above that agile requires a clearly defined business vision and direction.

"You are limiting the possibilities of Agile teams," I imagine my agilist friend, Kelly, say to me. "Agile is about building small units of functionality that provide empirical evidence for changing direction and adjusting your goals according to your findings. This is what being agile is all about."

I completely agree with Kelly, but we are really talking about a matter of decision scope. Software agile processes have been so successful because they help bring business process practitioners and technology implementers together to work on small bits of functionality and run experiments that can optimize and evolve a business approach to a stated goal. One of the biggest complaints that agile development teams have is that the business "does not know what they want." The agile process itself cannot define that business direction, though it can improve upon ways to achieve that direction once it is given. I have seen business leaders attempt to practice agile at higher levels of business leadership several times, only to be frustrated by it. That is because business leadership at the senior levels is about setting direction and creating a vision. It is very difficult to perform those tasks within a set of processes that were created as optimization practices, not creative practices.

Three examples of creativity will help us understand how process improvements, whether they are agile or AI, can provide great improvement when moving toward a given direction but cannot themselves provide the level of creativity needed to generate wholly new products and techniques. Two of our examples come from the familiar world of business, and the last one comes from the world of sports.

Microsoft and the Innovator's Dilemma

In the late 1990s and 2000s, Microsoft clearly dominated the world of end user software. The Microsoft Windows platform ran almost ubiquitously in all businesses and homes, leaving its competitors Apple and Unix in the dust. Even though Windows has lost much of its market share in the home PC market today, Microsoft continues to dominate in the world of desktop operating systems. While Microsoft Windows is still the clear leader in desktop operating systems, the computer desktop itself has begun to wane in popularity, causing a market shift in the last 20 years to move toward end user computing in the form of non-PC-based products.

In his book, *The Innovator's Dilemma*, Clayton Christiansen provides a number of examples where businesses that find high success in a particular product or service will often find themselves in a position of obsolescence due in large part to their own success. Christiansen's argument is that the more successful a company's product is, the more incentivized they are to optimize profits for that product. As they continue to increase market share and optimize profitability, outside factors, partially created as a side effect to their success, will move downmarket and begin to eat away at the product's user base. These products will often be specialized and offer lower capabilities but at a much lower price point. The company has optimized its own product for growth, which means new features, higher price points, and upmarket clientele. By the time the company has identified its lower cost and lower featured competitors, it is too late to make the shift in its product direction, and they are eventually strangled out of the market.

While Microsoft Windows was by no means strangled out of the marketplace, we can see the shift that took place in the consumer computing space through Google and Apple in the 2010s. Google created its own web browser with the intent of driving more traffic to its search engine, which generated income not on its own sales but through its advertising partners using the Google search engine. Google began to treat its Chrome browser as a fully functional computing environment, eventually releasing low-cost Chromebooks, which only needed light hardware and no need for the functionality of a heaving operating system such as Microsoft Windows.

In a world of data-driven decision-making, Microsoft's continued efforts to make a better operating system made a lot of sense. AI can help these kinds of optimization efforts to make an existing product more profitable in its operations and fully featured in its capabilities. In this case, the product vision, that is, the rules of the game, are clear–make Windows better. When a product like the Chromebook is introduced, however, it questions the very need for an optimized and fully functional operating system. In Christiansen's terms, AI processes will continue to optimize and make an excellent product, even after consumer demand has plummeted because it doesn't understand that the game has changed.

Apple's Revolutionary Product

Like a few of you technology enthusiasts out there, I have watched Steve Jobs' keynote speech introducing the iPhone from 2007 more than a couple of times. It was a fantastic example of what Clayton Christiansen called disruptive technology. There were a number of innovations that needed to be pushed in order to realize the iPhone vision, including computer hardware that was small enough to fit in a pocket, yet powerful enough to run the OS X. Just as important, the touchscreen technology needed to advance enough to allow for a digital keypad to replace the clunky and static plastic keyboards of Nokia and Blackberry. These two innovations were not obviously present in the sales data and customer sentiment surveys up to that point. The disruptive element of the iPhone was its convenience and portability factors applied to computer and networking functionality. From the perspective of established technology companies such as IBM and Microsoft, the hardware and software capabilities provided by the iPhone were extremely limited. Over time, as that direction of convenience and portability in computing devices has been established, Apple's smaller devices such as the iPhone and iPad have largely eclipsed the need for desktop computers in most of the ways consumers use these tools.

The Changing Game of Baseball

As a kid, I spent a lot of time listening to the 1980s Cardinals baseball dynasty on the radio. The hall of fame shortstop, Ozzie Smith, was my hero, and I remember many special occasions of going to his St. Louis restaurant, hoping to meet him (sadly, I never did). For college, I went north to the big city–Chicago. I fell in love with Chi-town and claimed the Cubs as my new hometown team, and they have been ever since.

Baseball has undergone several transformations over the century and a half since its inception, and the changes were not at all intuitive. One need look no further than the great baseball legend, Babe Ruth, to see how the sport is played completely different today than it was in the 1920s. Babe Ruth was an all-around great player–he was an amazing slugger, and his record of 714 career home runs stood undefeated for 50 years, when Hank Aaron beat that record with his 715th career home run in 1974. Twenty-five years later, in the 1990s and early 2000s, a new breed of home run powerhouses was knocking balls into the stratosphere. Players such as Barry Bonds, Sammy Sosa, and Mark McGuire were regularly breaking long-held home run records. Since that time, home runs have been a staple of major league baseball, and Babe Ruth, the once "Home Run King," fell from the top seat in the lists of home run leaders.

Several things happened to bring about this change. The first striking example is in the way these players actually practiced their sport. If one were to compare the physiology of Babe Ruth, who was known to down a beer and a couple of hot dogs between innings, with that of Barry Bonds, you would think they were trained for two completely different sports. Ruth had an intimidating build, no doubt, and used his size to power the ball over the fences. Barry Bonds, however, at his peak looked more like a bodybuilder than a baseball player. His contemporaries like Sammy Sosa and Mark McGuire also followed this same muscle-bound build. It's interesting to note that Babe Ruth played the position of pitcher. In modern baseball, pitchers and sluggers are trained much differently. They are specialized. This specialization did not exist in Babe Ruth's time. Players like Barry Bonds look like they are specifically built to hit home runs because they are. Training regiments and specialized diets make up a huge component to modern baseball.

The game's focus on specialization comes as a result of greater technology in terms of health science and equipment but also is a reaction to the advent of televised games. People watching the games at home appreciated the big moments in baseball that they saw through dramatic strikeouts from the pitchers and big home run moments from famous sluggers. The effect of television and its impact on the game was not something that could be found in the data from the game itself but from the periphery–from the television networks that increasingly became the primary sources of funding for major league baseball.

Conclusion: The Limits of AI-Driven Decisions

In each of these scenarios, the rules governing the domain in question are fundamentally changed. AI-driven decisions and insights work well within the confines of the game they are playing, and the information sets that are fed into the AI act as a kind of rulebook for the game. Within this set of rules, AI has the potential to find numerous opportunities for optimization and evolutionary enhancements, but it is up to humans to understand the domain at a high level and define the boundaries and rules of the game.

AI is very good at wading through massive amounts of data. In our baseball example, it is quite possible that an AI armed with statistical information from multiple players and teams could accurately judge which players will be top performing in the upcoming season. AI would be less capable, however, of anticipating longer scale aspects of the game that may be dependent on baseball's context within the larger world. For instance, certain rule changes such as the pitch clock have been implemented due to baseball's waning popularity in television statistics due to its long game time. Television ratings are not part of a core set of baseball data, though it is quite possible to include that data into some experimentation on the part of human handlers of the AI system, essentially adjusting the game rules of the AI reasoner. Innovators and entrepreneurs are able to see these possible connections between disparate knowledge domains and create new products and services that create new markets. In this way, AI acts as a force multiplier for innovative entrepreneurs and business leaders.

In the next chapter, we will be talking more about how we can use the concept of meta systems to understand the limitations of AI reasoning and utilize that information to our advantage when designing Enterprise AI applications.

CHAPTER 2

Meta Systems

Overview

We will begin this chapter with a discussion of Enterprise AI application design from the perspective of meta gaming. In many ways, AI Application design has a lot in common with game design, with its incentive structures and rule books. We will discuss AI systems utilized in real-time chess, strategy games, and sports.

Next, we will discuss the use of proxies in measuring human competence and activity, with a discussion of incentive structures and how we can identify areas of overfitting and evolutionary movement in our system designs. Understanding the idea of proxy and incentive systems will help us think through our AI functional design at a deeper level.

We will finish this chapter with a discussion of meta systems and formal system design.

Our discussion of incentive structures in games and meta systems thinking presented in this chapter will not only help us design our own Enterprise AI applications but also compete in a marketplace increasingly dominated by these systems.

© Anton Cagle, Ahmed Mohamed Ceifelnasr Ahmed 2024
A. Cagle and A. M. C. Ahmed, *Architecting Enterprise AI Applications*,
https://doi.org/10.1007/979-8-8688-0902-6_2

Centaur Computing

Over 2 years in 1996 and 1997, Garry Kasparov, the world chess grandmaster at the time, played a series of championship chess tournaments against IBM's Big Blue supercomputer. The media billed this competition as humanity's last stand against the machine. Kasperov handily won the first tournament in 1996, beating Big Blue 4 to 2. In 1997, Kasperov was forced to resign in less than 20 moves after a 2 ½ to 2 ½ tournament tie. This was hailed as a major turning point in the evolution of artificial intelligence.

After a year of deep introspection, Kasperov returned to the world of chess with new-found insights from his years of playing against the supercomputer. Kasperov introduced a new form of chess competition which he called Centaur Chess, otherwise known as Advanced Chess. This is a form of chess in which two teams each play a tournament set of chess games against each other. Each team consists of one human and one AI engine. Since its inception, Centaur Chess teams have proven to outperform both single chess AI engines and individual grandmasters in tournament games.

Kasperov's strategy against Big Blue was to rely on long-term strategy over aggressive play. Big Blue and other chess engines tended to play very materialistically, that is, they would place a heavy value on taking material from the board opportunistically. Kasperov took advantage of this tendency by playing out very long strategies, or playing defensively, hoping that the AI would eventually play itself into a corner.

The fatal flaw that human chess players have is the tendency to make small errors. The AI is consistent, if not necessarily inventive. In Centaur Chess, the human captain (or captains) can focus on long-term strategic direction while utilizing the AI engines to check for any blunders before proceeding. Since that time, chess engines have increased in intelligence and can easily run on our smartphone. Centaur teams–teams that consist of the brain and imagination of a human, coupled with the horsepower of a computer–continue to beat even the most advanced AI engines by themselves.

Age of Empires and Hacking the AI

It comes as no surprise that elite players of Centaur Chess are also avid players of other games. It has been rumored that certain elite players are also top players of real-time strategy games, particularly Age of Empires. Games such as Age of Empires (AOE) require long-term strategic thinking and quick moves in order to compete at elite levels. AOE is also known for its innovative AI opposition.

In games such as chess and AOE, human players must use long-term strategies to their advantage. The "materialistic" tendencies Kasperov noted in his early games against Big Blue are also seen in AOE AI opponents.

Years ago when I first began playing AOE, I found that it was easy to draw out the enemy AI by building large empty structures of complicated walls near the enemy territory. The walls themselves have no real value to humans, except in terms of protecting high-value buildings and personnel. In the first iteration of the game, one particularly enjoyable hack against the AI was to create elaborate labyrinths near the entrance of your base and drawing enemy armies into them. They would soon get trapped, and you could more easily pick them off one by one as they emerged from the labyrinth.

Newer versions of the game have since fixed the labyrinth hack, but up until recently it was still possible to randomly place sections of wall along the path between your base and the AI opponent's base as diversions along the way. The AI would almost invariably stop to destroy the random pieces of wall in their path without just going around them. This is a classic example of materialistic incentives taking the place of higher level strategy. Typically, RTS AI opponent engines have tried to make up for this limitation through giving the AI an unfair speed advantage by allowing their units to move much faster than a human player.

21

The AI Goalkeeper

As I am writing this, there are a number of online videos being released showing pro footballers going up against an AI-powered goalkeeper. The AI anticipates the shots based on color tracking. Speed does not work well against this bot as its main advantage is the speed of its movements against an oncoming ball. When the players first go up against the AI, it seems unbeatable, as indeed it does have superhuman speed and reflexes. By making several unsuccessful shots on goal, the players start to figure out that it has a hard time reading trick shots with spin, and it is possible to get some shots through by spinning the ball in a way that moves the ball away from the goalkeeper at the last minute. This of course takes a lot of skill by the player to do this.

There was another approach to this robokeeper that seemed to require less ball-handling skill and more human ingenuity. AI's own speed can be used against it if the player kicks the ball very slowly and causes it to bounce its way into the goal. It's humorous to watch this superhuman goalkeeper begin to fan ridiculously back and forth to anticipate the slowly bouncing ball, only to let it slowly bounce underneath it and into the goal. In a real match, this slow bouncing approach wouldn't work even with an AI goalie of course because if humans are assisting the AI, they can easily anticipate this slow ball and kick it away.

Marines vs. AI Patrol

Several years ago, the US military introduced its first prototype of an AI patroller bot that identified militants as they approached a target. A group of eight marines were tasked with finding a way to infiltrate the position guarded by the AI patrol. On first pass, it seemed there is no way to hide from this AI guard as it was trained on all forms of human movement and body patterns. It did not take long for this set of marines to figure out alternative ways to get around the patroller than just hiding behind trees. The most obvious way of getting around the bot was to use a strategy taken right out of the outlandish military series, *Metal Gear Solid*: hiding in a box. The marines found that they could walk undetected in plain sight as long as they covered themselves with a cardboard box. After that, the marines decided to have some fun with the bot by trying all sorts of nonstandard approaches, including doing somersaults all the way through the patrolled area.

It is in moments like these when the magic trick is revealed, and the young boy yells out that the emperor has no clothes. AI is excellent at seeing the patterns it is trained to see. Even as the AI gets better, and learns how to anticipate slowly bounced balls and somersaulting marines, the human ability to adapt will always give the AI a run for its money, leaving the AI always one step behind as it has to learn from human ingenuity. From these examples of games involving strategy, sport, and military tactics, we are learning over and over that the pairing of human and AI intelligence together generates amazing winning results that neither human nor AI could achieve on their own.

Human Competence and Its Proxies

In late 2022, there was a collective gasp in the tech community when Stanford researchers let the new GPT-4 LLM model loose on the bar exam, the standard exam administered to law school graduates required for becoming a lawyer. While the first iterations of the LLM were not able to pass the exam, GPT-4 not only passed but passed in the 90th percentile. "If lawyers can be replaced," the worried hand wringers in Silicon Valley asked, "what jobs will be next?"

Law is a complicated profession, requiring many years of schooling and on-the-job training. A good lawyer is very familiar with case law in their given expertise. But this is not the only qualification for a lawyer.

A lawyer must be able to win legal cases, often in front of a jury. The bar exam tests how well a would-be lawyer understands the technicalities of the legal profession, but this is only a fraction of what is needed to be a good lawyer.

Perhaps the most impressive use case of LLMs today is their ability to retrieve and summarize technical information of all sizes, from a single PDF to a century worth of case law. A law school student armed with a smartphone could easily retrieve the answers to the exam within seconds.

The same predicament is seen on a much broader scale in online high school and college courses across the world. In some ways, this makes these forms of testing obsolete, but it does not make the education system itself obsolete, it just means we must think about education in a different way.

Exams and certifications are, at their heart, a proxy for competence. If one has spent the necessary time and dedication learning the intricacies of case law, professional lawyer associations may generally assume that the person is on their way to meeting the stringent requirements needed to become a lawyer. The same is true of a high school or college degree or most professional certifications. The investment of time and attention over a certain time span (i.e., 4 years) generally required to pass a set of exams from memory alone has traditionally been seen as a strong indicator of a certain level of competence in a technical area. As AI reduces this time and attention factor, old forms of competence proxies are made obsolete, and new ones must be discovered.

This happens in business as well as education. This is seen in the checklist duties that many professionals perform as part of their everyday duties. Certain rather mundane duties are assigned to professionals as a way of "keeping the lights on." As automation is introduced, and those routines are disrupted or made obsolete, it is often discovered that the real value of a human performing such duties was not held in the mundane task itself that was automated but rather in the attention and familiarity with the environment that come with performing those routines, like a cop on the beat who drives around to regular checkpoints. It is not that those checkpoints are particularly problematic, indeed a regular police presence itself will clear up would-be problems in those areas. It is the familiarity of the turf that is gained by the police officer, and the places in between that are only attended to implicitly as they drive by, but enough to recognize almost subconsciously when something is "not right."

As we create our AI systems and automate large areas of our business, it is important to pay attention to this idea of competence proxying. One particular example comes to mind that is not specific to AI but is illustrative of my point. Many large companies still have a very large (usually overly large) change control board that meets several times per week. One of the first targets of eager young automation experts is to automate the creation of change control tickets and the approval process. This is not difficult, but in my experience, this approach has always been faulty on at least two levels. First, the idea of automating the creation and approval of tickets may be an easily attainable goal, but it misses the point of the change control process. This process is in place as a means of manual auditability. Rather than automating the ticket creation process, it would be better to build auditability right into the system itself, eliminating the need for ticket approvals altogether.

This automation brings about a second difficulty however. The change control meetings themselves were about auditability, yes, but they were also a forum for bringing employees together from across multiple teams to discuss the current activities in the technical organization. Once this process has been fully automated, and perhaps a series of emails or chats can take the place of regular meetings, general awareness of what is happening across the teams has a tendency to drop. This is a secondary function of those change control meetings, and another solution should be created to fill this newly created gap of human interaction and organizational awareness.

The Right Incentives

Incentive structures are key to AI systems as they give the AI measurable targets toward which they can aim. These incentive structures allow AI systems to learn and evolve over time. As we have seen earlier in this book, AI systems must be provided with an intrinsic incentive structure as it is not innate within these systems themselves. It is possible that an AI can create a series of localized goals in an effort to reach a higher goal, but the goals at the highest levels of the AI's intentionality must be provided by human handlers. For this reason, it is very important that AI application designers and handlers understand the appropriate incentives to direct their AI applications.

Overfitting

The most important danger that AI model designers must avoid is overfitting. Overfitting is the tendency of AI models to fashion themselves exactly after the test data that they are developed with. This test data is a proxy of real-world data, and by its very nature includes ancillary features that are not part of the system design. For instance, if a model trained to recognize certain types of trees is trained on pictures of trees in an urban environment, it is possible that this model could fail in production against pictures of trees that do not include cars, as cars are nearly ubiquitous in an urban environment. The model has recognized not only different tree types but has now assumed that all pictures of trees also include parked cars in the background because all the training data was taken within an urban landscape. The real danger is that the model developer sees the model achieving increasingly high scores as it is trained over and over on a single set of training data. Unbeknownst to the developer, the model is fitting itself to the training data–that is, trees in an urban environment along city roads–and not against the real purpose the developer had in mind, which is tree types themselves, regardless of environment.

Business process automation can also be overfitted in a similar way by incentivizing the AI applications on proxies and not underlying outcomes. Quality and efficiency programs such as Lean and Six Sigma have worked to simplify business process for many years and provide us some insights in this regard. While improvement initiatives in these methodologies do not address traditional AI and machine learning issues of overfitting, the concept of overfitting is useful when utilizing business process analysis tools such as Lean and Six Sigma for defining business process automation.

One of the core tenets of the Lean/Six Sigma methodology is to identify the existing business processes in place and then identify areas of waste within these systems. Waste in business process often comes in the form of extra steps or extra handoffs between business actors or systems. This waste can often be the result of proxying behavior. Auditing is a common example of this. Multiple code review steps and manual button pushing events are put in place in order to appease auditability concerns. In the case of code release processes in CI/CD flows, one of two things happens throughout the lifetime of these manual checkpoints. On one hand, the system changes may be managed down to such a minute level that a manual step becomes trivial and performs no real value other than a checkbox that a particular person or department representative saw the change. As these small changes increase in frequency, the process not only becomes a bottleneck but just becomes noise in the system, and the manual checks are just rubber stamps that are pushed along without any scrutiny. On the other end of the spectrum, if system changes are complex and involve several systems, it is likely that a single person or department cannot realistically be expected to understand the implications of the change that they are expected to approve.

In both of these cases, it is important to understand the full system context, thus not performing a one-to-one AI automation capability to replace these steps. It is likely that there are decision points further upstream in the process that can more efficiently satisfy the intent of the manual processes. In some cases where the manual processes are themselves small proxies for a separate series of undocumented processes, whole new applications may spin out of a small proxy step.

The coupling of AI practices along with process improvement methodologies such as Lean/Six Sigma can yield very successful results, particularly when designing your Enterprise AI applications as successors to existing semimanual enterprise process flows. Using these process methodologies can help identify and clarify business incentives inherent in each process step. This is especially important for AI applications as automating poor incentive structures is a much higher risk because of their tendency to massively scale and run without human intermediators who may otherwise find workarounds outside of the system.

Formal Systems and Meta Systems

In his Pulitzer Prize winning book, *Godel, Escher, Bach: An Eternal Golden Braid*, Douglas Hofstadter uses a formal system called the MU puzzle to exemplify how AI operates within a formal system and how human designers have the ability to jump outside of the system and think at the meta level. The book in which this puzzle is presented is fully deserving of its Pulitzer Prize, and I would encourage you to pick it up. As such, I won't present the full puzzle here in anticipation of your extracurricular reading assignment, but I will skim over the main points of his puzzle.

The MU puzzle is presented as a series of simple rules applied to a starting string ("MIU") with the goal of achieving a specified string output ("MU"). There are a handful of rules that add, remove, and replace the "M," "I," and "U" characters, with the ultimate goal of producing the string "MU" from the starting axiom of "MIU." The puzzle itself is unsolvable given the starting axiom ("MIU") and the limited ruleset. The unsolvability of this puzzle is not explicitly obvious, and only through trial and error does the player begin to suspect that there is no solution. Hofstadter's exercise illustrates the idea that in order to solve the puzzle, the player must jump outside of the rules of the game itself and begin questioning those rules, not as a player of the game but as a designer of the system. An AI that has been programmed to solve this puzzle first of all has no explicit indicators that this puzzle is unsolvable and could go on adding more and more characters and subtracting them in endless combinations *ad infinitum.* The human player begins to intuit that there is something wrong with the game and doesn't have to play too long before they start recognizing some broad patterns in their string combinations that hint to the puzzle's unsolvability.

The book was written during the early days of chess engines, and he discusses a surprising chess engine that showed some early promise of being able to recognize these kinds of high-level patterns. This particular chess engine was developed in Canada and by all objective measures of game play was the least successful of its competitors. Its developers discovered, however, that it had a tendency to forfeit very early in game play, as soon as it discovered that it was in an unwinnable situation. This was fascinating because while the engine itself could not figure out how to win the game, it could interpret enough patterns about the game itself to see that it was in an unwinnable position and stop playing. In Hofstadter's terminology, the Canadian chess engine would "jump out of the system" by abruptly quitting the game.

Enterprise AI applications require mechanisms to provide the capabilities to "jump out of the system." Throughout this book, we will see examples of AI systems (we call them agents) assigned to watch other AI systems, acting as a "meta system." In this way, Enterprise AI applications are really sets of interlocking, conversational, and sometimes adversarial AI systems that are built with different incentive structures. Systems such as Process Anomaly Detection can work as observability agents that keep other processes in check. While it is possible to build all of the competing incentives into a single mother brain, this has proven historically difficult and practically very difficult to maintain. On the other extreme, the proliferation of agents performing tasks coupled with watchers of those tasks can become overly burdensome if the tasks themselves are not carefully considered in terms of their real business value and possibly irrelevant proxy value as a business competence indicator.

Conclusion

In this chapter, we have discussed the concept of meta systems in relation to gaming, proxying, business process, and formal systems. This chapter has provided a high-level overview and introduction to many important concepts that can be researched independently. Particularly, the concept of formal systems, as described throughout the works of Douglas Hofstadter, is very useful in designing Enterprise AI systems. As we look to automate and modernize many of our existing business processes, methodologies such as Lean and Six Sigma can prove invaluable to designers as they look to replace inefficient and semimanual processes with AI capabilities.

In the next chapter, we will discuss the predictive aspects of AI and how an understanding of certain AI agents as "prediction machines" can help us both understand how these algorithms work and identify new opportunities for providing customer value.

CHAPTER 3

Prediction Machines

Introduction

In 2017, a small group of Harvard researchers (Ajay Agrawal, Joshua Gans, and Avi Goldfarb) released a book called *Prediction Machines*, in which they proposed that the best way to discover AI application use cases is to look for problems of prediction. AI is, at its heart, a "prediction machine" that can be used to find correlations and patterns in past events to predict future outcomes.

Framing AI applications as prediction machines is useful for both understanding how specific AI implementations such as generative AI work under the covers and also in our identification of new opportunities for implementing successful Enterprise AI applications.

In this chapter, we will explore the nature of AI applications as "prediction machines" and contrast it with our various understandings of human creativity. This framing will give us the right lens from which to view our enterprise use cases and help us design systems that optimally harmonize both AI prediction and human ingenuity.

Prediction in AI

My grandparents were married for over 50 years, and after that time, together, they were well acquainted with each other's habits, interests, old stories, opinions, and speech patterns. Most of us have known an old couple who can "finish each other's sentences," as the old saying goes. While living with someone for 50 years will definitely give you the ability to predict your partner's behavior, we are pretty good at understanding people after a much shorter time. Part of the stress of getting a new job is not being able to anticipate what your boss wants. After some time at the job–perhaps a quarter, perhaps a year–we begin to understand how our boss thinks, what drives them, and pick up on signals from our boss that indicate whether or not they are pleased with the direction each of our tasks is taking.

At a more public level, many comedians are excellent at picking up behavior patterns of certain well-known people and doing impressions of them based on a careful observation of their public work such as movies, television shows, and interviews. Some public personalities have much more exaggerated and unique personalities than others and can be impersonated easily (Elvis Presley, Bill Clinton, etc.). Impersonations can be seen as a kind of prediction as we are imagining the behavior and speech patterns of a particular person in a novel situation.

At the broadest level, we use prediction in our daily social interactions as we have certain social conventions and speech patterns that help us work with others every day, whether we are driving on the freeway or exchanging pleasantries with a cashier at the grocery store. This can be understood most readily through listening to American top 40 pop music radio. The morning show radio DJs on these stations are locally recognizable (and usually the highest paid local radio announcers), and they speak in comfortable cadences with mostly familiar patterns of speech. The reason for this is to help their listeners wake up and enjoy their commute to work. Songs that successfully hit the top 40 charts often make strategic use of common phrases and popular cliches sprinkled into their song lyrics. This helps the listener quickly anticipate and familiarize themselves with the melody and rhyme of the song. This is often what makes these songs catchy, and, of course, predictable. This is also what makes some of these songs short-lived, as that predictability usually wears thin on the listener after a while.

All these examples of human social predictivity can help us understand how AI prediction works. Generative AI models are fed a huge amount of information in the form of training data and then asked to complete a pattern based on this input.

Imagine you're at a party. The music's loud and conversations buzz around you. You're talking to someone you have just been introduced to, and they are fascinating. Then, mid-sentence, the DJ cranks the volume of the music up and plays a banger of a song. Everyone gets into it, that is, except you and your new-found companion. Your new friend asks you a question, and you can only make out the first part of it. You are afraid of embarrassing yourself, but lucky for you, humans are really good at understanding each other, so you are able to make a good guess as to the completion of your question, and the conversation goes on as normal–outside perhaps–with your new friend. Generative AI works in a similar way to generate sentences based on a prompt–it considers the context of the audience, the persona it is impersonating, and the corpus of language it has ingested to predict the words to string together.

Generative AI, such as the well-known GPT models, operates by predicting the next word based on the words that came before. It uses context-based pattern recognition, constantly calculating probabilities. If you start a sentence with "The cat sat on the," it might predict "mat" because, statistically, that's a common phrase.

Tuning the Prediction Engine: Hyperparameters

For our discussion of AI hyperparameters, let's use the analogy of driving a car in which you have control over the speed, the direction, and even the comfort of the ride. Similarly, in the world of AI, hyperparameters are the knobs and dials that control how the AI makes its predictions. One does not expect a minivan to handle in the same way as a race car, though both will get you from point A to point B. Managing hyperparameters can help provide the appropriate expectations for your ride along the way.

Popular GPT models work on three basic hyperparameters.

Temperature

A low temperature (closer to 0) makes the AI conservative, sticking to safe, predictable words. High temperature (closer to 1) makes it adventurous, generating more diverse and creative outputs. The temperature hyperparameter affects the randomness of predictions made by an AI model.

Top-k sampling

Top-k sampling restricts the AI to a fixed number of probable words (k). If k is 50, it only considers the top 50 most likely next words, balancing predictability and variety. Top-k sampling restricts AI's choices to a fixed number of the most probable next words (or tokens). You can directly control the diversity by adjusting k. A lower k keeps the predictions more conservative, while a higher k allows more variability. The main benefit of this hyperparameter is that it is easy to implement and understand. Depending on the context, the fixed value of k can sometimes be too restrictive or too lenient based on the probability range between the k-th and (k+1)-th words, leading to suboptimal choices.

Top-p (nucleus) Sampling

Top-p sampling considers the smallest set of words whose cumulative probability exceeds a threshold p. If p is 0.9, it means the AI picks from the top 90% probable words, making the output both coherent and interesting. The threshold p adjusts dynamically to the probability distribution. This means that if there are few highly probable words, the selection will be narrow, but if there are many words with similar probabilities, the selection will be broader. By focusing on cumulative probability, top-p sampling can balance the need for diversity and coherence better than top-k in many cases.

Understanding these sampling methods helps business managers fine-tune AI outputs for various applications. For tasks requiring high reliability and consistency, like generating legal documents or running production jobs, use lower values of k or p and a lower temperature. For creative output, such as marketing copy, some forms of user documentation, and conversational bots, use higher values of k and p and a higher temperature to encourage conversational, natural, and more engaging outputs.

Beyond Words: Predicting Images

Generative AI for image creation works in the same way, but instead of predicting words, it is predicting pixels. Image generation models like DALL-E use vast datasets of images and corresponding descriptions. The model learns to generate new images by predicting pixel patterns that match a given description.

Here, hyperparameters control the style, coherence, and novelty of the images. For instance, you can guide the AI to create images that are more realistic or more abstract, depending on your business needs. The goal remains the same: to predict the next element in a sequence, be it a word or a pixel, with the right balance of creativity and accuracy.

Machine Learning: Training the Prediction Engine

Machine learning is about teaching a computer to recognize patterns and make predictions based on those patterns. There are various types of machine learning algorithms, each using prediction in distinct ways to train models.

Supervised Learning: I still randomly find old flashcards of "100 First Words" scattered among my board games from when my children were very small. Many of us parents are familiar with these cards–a plain white, oversized card with a picture of a floating red ball in the center and the caption "ball" along the bottom. This labeling of images with words is a form of supervised learning. The AI is given a set of inputs and the corresponding correct outputs. It learns to predict the output (the word "ball") from the input (a picture of the red ball) over time through high iterations of trial and error, minimizing errors over time.

Unsupervised Learning: I studied ancient Greek for 4 years in college, and as a cautionary tale in my first semester we were told a story of a young man a couple hundred years ago who found a copy of the Bible in its original Greek. He sat under a tree, so the story goes, and taught himself to read ancient Greek from this single text. The people of this rural American town were incredulous and cast him out of the town because they believed that ancient Greek was so difficult to learn that there was no way a young man could learn it on his own without a tutor (or without some nefarious supernatural help). Unsupervised learning works without the help of any tutor, human, or otherwise. Here, the AI is given data without explicit instructions on what to do with it. It predicts patterns and structures within the data until it can repeat these patterns in both novel and understandable ways.

Semisupervised Learning: This is a blend of the above two methods, using a small amount of labeled data to guide predictions on a larger set of unlabeled data, akin to using a partial map to navigate a new city.

Reinforcement Learning: This is the well-known behaviorist approach to learning, as exemplified by Pavlov's Dog. The AI takes actions in an environment and receives feedback in the form of rewards or penalties. It learns to predict which actions will maximize rewards over time, refining its strategy through trial and error.

Human Prediction and Creativity

Human ingenuity includes a mix of both prediction and creativity. The predictions we make are based on an incalculable amount of inputs we receive that color our predictions around how our fellow humans will act and how particular decisions in our personal fields of expertise will play out. AI, at least for any foreseeable future, suffers from a lack of input. This lack of input comes in two forms, implicit and explicit.

Implicit Input Models

Implicit input models are best seen in "big data" applications. Throughout the 2010s, tech companies scrambled to collect as much data as possible. "Data is the new oil" was the key phrase. While that direction has not gone away, it will always be limited to the number of input sensors that can collect data. The old phrase "what gets measured gets done" can be easily rephrased as "what gets measured gets known," and the known world of AI is still very small. With governments, corporations, and individuals actively attempting to limit the amount of data collected and shared with corporate AI systems, this problem for AI is only getting worse.

While implicit big data models are still not better than humans at making expert decisions, they have provided two significant advances.

First, unsupervised machine learning models fed on huge datasets have consistently been able to show us hidden patterns and features in our data, challenging our understanding of the information models with which we work. For instance, large data models in corporate operational systems can often draw correlations between disparate events that may help system engineers discover underlying issues that were previously unconnected.

The second big advancement in implicit big data models is the rise of large language models. While there is legitimate concern over these models' ability to accurately (and ethically) provide answers to questions about the ingested data, researchers have discovered that the more data ingested by our current LLM systems, the better they are at understanding human language. In a very tongue-in-cheek way, I have described that usefulness of public LLMs as getting their super ability to understand English through ingesting a decade worth of Reddit data. While this has proven an excellent way to learn English, you would not be quick to ask this same bot what its opinions of a given matter that it has learned from the School of Reddit.

Explicit Input Models

In many ways, the implicit models provided by the LLMs have given a huge boost to custom AI applications that can make excellent use of smaller explicit data models. Businesses can now utilize these LLMs, that have been trained on billions of data points, and train them on very specific information sets, such as their own technical documentation, custom application operations data, and even streaming corporate communications such as email and chat. Since LLMs have such a strong understanding of English, they can more easily understand and interpret smaller sets of explicit input, such as a single pdf.

The limitation of data input can act as a way to purposely focus an AI application's output on what is important and relevant to its purpose. In this way, limiting the scope and range of predictions within your AI application can be extremely useful; in fact, we heavily encourage that approach throughout this book. Small, quality information sets can be used and provide excellent prediction models for these limited domains.

Just as there was a downside to implicit input models in the form of limited models, explicit models also suffer from issues of measurement.

The problem of explicitly choosing the right inputs for your AI models is similar to giving and taking good advice. In his book *The Black Swan*, economist Nicolas Nassim Taleb describes how he chose and worked with his personal trainer. He chose someone at his gym who was someone close to his age (middle aged) and looked the way he wanted to look (like a weightlifter). Taleb then paid this man to let Taleb follow him around the gym when the trainer worked out. The trainer provided a lot of advice and instruction to Taleb, but as he points out in his book, more often than not this advice did not actually correspond to what the trainer did in his own personal training regiments. Taleb would often ask him "do you follow this advice yourself?" to which the answer was invariable, "well, no but…"

The point Taleb is making in this story (and throughout his book which is basically about applying extreme empirical thinking to economics) is that the advice we give, even when we think it is good advice, does not always follow the actual actions we take. We create stories for ourselves in our minds that do not always follow the empirical realities of our day-to-day activities.

When setting up your explicit measurements for your AI applications, it is important to keep this in mind, because the narrowing of focus could become a detriment to us if we do not fully understand our problem domain, or act primarily on ideals and theory rather than empirical data.

In building large Enterprise AI systems, it is good to blend both of these models. Implicit models are excellent for discovery and challenging our assumptions about our understanding of a complex and evolving environment, while explicit systems work very well at minimizing hallucinations, operating efficiently, and maintaining supportability over time.

Gen AI–Generative and General

There is a significant difference in the use of the phrase "Gen AI." Before 2022, the phrase was broadly used by the research and engineering communities as "General Artificial Intelligence." Articles written about AI up to that point would talk about Gen AI as a future that was just around the corner. Once ChatGPT became mainstream, the phrase Gen AI became the new buzzword and referred to the new capabilities of generative artificial intelligence. Generative AI's ability to "generate" new content is based on the statistical predictive capabilities we have been discussing in this chapter. This ability to generate predictions based on inputs, however, does not imply the capability of a general intelligence.

When AI professionals discuss general intelligence, they are referring to an intelligent system that has broad general knowledge about the world to the point that it can make human-like decisions and form independent conclusions based on a general skill in critical reasoning. AI as we know it today, and especially the new popular flavor of generative AI, creates predictions based on existing inputted information. AI is not creative in the way that we think about creativity as humans, as we do not solely extrapolate existing historical patterns but drive toward creating a future that often looks very different from the past. This is the sole of business innovation and entrepreneurship.

A number of years ago, I saw an example of this at a small ecommerce company I was working at (some of the details in this story have been changed to protect the innocent). One Monday morning, my team received a concerned call from our marketing department. They said that they were seriously concerned about the sales on our website, as they were significantly lower over the weekend than expected. We checked all of our monitoring tools and saw no notifications of lower sales. We had observability tools in place that track sales, customer traffic, bot traffic, customer journey flows, average cart size, and many other metrics that determine an acceptable amount of traffic, and whenever there was an anomaly in those metrics, we would get alerted. We assured them that our web traffic was consistent with what we had seen the previous couple of weeks. They were adamant and said they expected a significant increase as they had released three marketing campaigns over the weekend and were not seeing the kind of sales we were expecting.

Unfortunately, no one alerted the technical team of these new campaigns, otherwise we could have been looking for the change in traffic. As it was, we were using machine learning anomaly detectors to alert us to any changes that broke historic patterns. When changes did not happen, the algorithms assumed everything was just fine. In this case, a new process needed to be created that did not just take the existing historical data into account for anomaly alerting but take the human predictions from the marketing department on the expected outcomes of the new promotional campaigns into account as well. In the case of most AI applications, the algorithms anticipate the status quo and will only adjust if the algorithms are manually adjusted.

The job of business leadership is to either adjust to evolving demand in an existing market or create new markets for opportunity. In both cases, business leaders are creating new approaches not based solely on historical data but an idea of a future that has not yet come into being. The vision of this future is complex and based on an understanding of how humans will react in the marketplace (i.e., what products and services they will engage with based on their evolving wants and needs). To assume that AI will be able to anticipate human decisions if just given enough inputs to enable a general reasoning capability is similar to an economist assuming that they can make accurate market predictions based on an understanding of what is best for people in the market. What we have discovered in the past several decades, however, is that people do not always make decisions based on what is best for them (people are irrational actors, as economists say), and AI has not been able to reliably predict human behavior at the market level.

In essence, while AI prediction engines can simulate aspects of human creativity, they do so through patterns and probabilities. Human creativity, which is based in large part on empathy and unexpected serendipity, offers a contrast that highlights both the potential and the limitations of AI as a prediction machine.

Problems of Prediction

We started this chapter with a reference to the book *Prediction Machines*. The core example of a prediction problem that the authors present is the airport lounge. The airport lounge exists as a waiting room for travelers who are waiting for their flight. Airport designers and administrators do their best to make the area comfortable for travelers who are waiting for a delayed flight or temporarily stopping by during a layover.

These waiting areas, no matter how cozy they may be, only exist due to the fact that flights could not be connected efficiently, and the travelers were not lined up appropriately to the timing of their flights. The problem of getting a traveler from their hotel room to the airport, onto the flight, and then to their ultimate destination is complex, but a vast improvement on this situation could be made with the right coordination of AI tools and human ingenuity. If airports and airlines were more interconnected with each other, connected to individual travelers via their phones, and connected to the means of travel to and from the airport, the downtime travelers experience would be greatly reduced. Beyond the system integration, though, the real problem at hand is predicting when there will be a delay in travel and when there will be a mismatch between the time the traveler gets to the airport and the time they board the plane. The human capacity to provide this information manually is not really possible. AI is able to ingest and integrate real-time information regarding traffic patterns, weather, plane schedules, and other such information and predict dynamic arrival and departure times down to the minute, that is, of course, assuming that the AI has interconnected access to all of this input.

Non-Places and Non-Events

French Urban Theorist Marc Auge coined the term "non-places" to describe places like the airport lounge. Non-places are areas that we move through regularly but have no history or purpose beyond this "just passing through" phase. They are not spaces where people meet, such as parks or a local cafe–they have no history or social purpose of either bringing people together to work or to provide a particularly relevant place for individual effort. These non-places are just fillers in our cities, and our modern urban landscape is full of them–subway undergrounds, highways, bus depots, and even large swaths of city sidewalks along monotonous city blocks.

A non-event is similar to a non-place in that it does not in itself serve a meaningful purpose other than a pure transitory or transactional role. The most obvious non-event is a fast-food checkout counter. Kiosks and phone apps are widely available today, but they are extremely rudimentary compared with their human counterparts at the cash register. New advances in AI vision and speech recognition will make these experiences virtually unrecognizable from the clumsy kiosks of today.

Our experience with these transitory places provides excellent opportunities for introducing AI-powered technology to alter our experience of these places or to eliminate their need altogether. Retail and ecommerce have made massive strides in this area as recommender systems use prediction to help customers quickly find everyday items they regularly purchase or will likely need, eliminating the need for the tediousness of buying staples such as laundry soap, kitchen ingredients, and other daily essentials.

AI has played a huge role in creating great products that predict our consumer needs, plan our travel, and manage our weekly calendars. Two ways of looking for opportunities in these areas are to look for places where humans can be either eliminated or repurposed from routine tasks. In the case of self-driving cars, for instance, the human need to perform the complex and dangerous task of driving a car on the highway can be at least partially relegated to an AI driving assistant. The voyage may take the same amount of time, but now the human's experience of the journey has changed since they do not have to focus quite exclusively on the task of driving.

Similarly, cash register operators or restaurant waiters are not just displaced by AI ordering mechanisms but can now be repurposed to provide a better customer experience. For instance, a waiter who no longer has to remember and record individual orders at each table or learn how to enter orders into a system will be able to shift their focus to conversation and friendly interactions that will make the dining experience much more enjoyable to the customer.

The initial reactions to AI around employee displacement is an unfortunate part of the technology cycle. As AI becomes more commodity over time, human employees will be brought in to perform duties that are more suited to human conversation, interaction, and overall system management and handling.

Conclusion

In this chapter, we have discussed the idea of AI applications as prediction machines. We started by describing some of the technical details involved in AI prediction-making algorithms including hyperparameters and the learning models that shape the accuracy of their predictions. We then contrasted these prediction-making algorithms to an understanding of human ingenuity. We ended this chapter by exploring possible avenues for AI application opportunities through the lens of prediction problems, non-places, and non-events.

In Part 2, we will dig into the technical architecture of Enterprise AI applications as we explore agents, machine learning systems, and other technical aspects around building your Enterprise AI application.

PART II

Designing Your AI Application

CHAPTER 4

Anatomy of an AI Application

In this chapter, we will lay out an architectural framework for AI-based software.

We will begin with the story of the evolutionary progress of modern AI applications in the enterprise and how AI applications differ in their approach to application development from traditional applications.

Business process is at the center of Enterprise AI applications, so we will spend some time discussing business process concepts and business ontologies.

Next, we will discuss AI agents. We will describe what they are, the different types of agents, and how they work together.

Finally, we will discuss the AI application stack. We will discuss how each of the different components of the AI application stack work together to create robust and maintainable enterprise solutions.

The Evolution of Enterprise Software Development

There is a fine distinction between AI applications and what I am calling "traditional" applications. At the highest level, all software is a means of solving real-world problems, and in a business setting those problems

A. Cagle and A. M. C. Ahmed, *Architecting Enterprise AI Applications*,
https://doi.org/10.1007/979-8-8688-0902-6_4

have to do with solving particular use cases that are involved in running a business. Useful software is always built to solve problems of the users, and in that sense there is no real difference between an application that is written as a mainframe process, a website, or an AI application.

At the broadest level, AI is just the next level of automation in software, and its nature does not really change just because we have the latest set of programming tools and compute platforms.

Abstraction at this highest level is not really that useful to our discussion. We need to dig down a level deeper to the actual ways and means of building the software to understand the advantages and game-changing nature of building AI applications.

AI applications are evolutionary in nature, but they represent a massive jump in the progress of building software that is comparable to the birth of Internet-based distributed systems of the late 1990s. Before distributed systems, and particularly before the wide-scale adoption of the Internet for ecommerce, software was relegated to very large, complex, and expensive monolithic systems. Programming on these systems was difficult, and it was hard to learn as one needed to have access to one of these systems through a college or business to practice and learn. With the advent of highly distributed applications, particularly through web development, teenagers and interested adults could now teach themselves programming from home and run their applications on fairly inexpensive hardware. This marked a level of democratization in software development that allowed software applications to be written by anyone who had an engineer's patience, methodical approach, and capacity for logic.

Traditional applications of today generally consist of several programming languages and data platforms. In the early 2000s, this was broken up into a three-tier architecture that typically included a database tier, business logic tier, and presentation tier. These systems were large and costly, and businesses typically would invest in a small number of enterprise technology stacks. Microsoft was particularly successful in this area, providing a stack that included SQL Server (database), C# (business logic), and ASP (presentation tier) that worked together fairly well. In the open-source community, the LAMP stack ((L)inux, (A)pache, (M)ySQL, (P)HP) was adopted as a standard for software tier interoperability. The reason for these technology stacks was not only for interoperability but also for support. It took an engineer a long time to become proficient in any one of these programming languages or data platforms. While the algorithmic and data structures were based on similar foundations, the programming syntax and hardware platforms were different enough that they made the skillcap for any one technology platform fairly high. This was also partially due to the lack of training materials available in the early 2000s. Like many of my fellow engineers at that time, I was very proud of my work library that was full of very thick programming manuals that usually had animal mascots prominently displayed on their covers.

As these applications grew in size and complexity, enterprises began focusing on reducing their ever-growing cost of maintenance and cost of change. From an organizational perspective, companies that built software began adopting Agile frameworks for organizing application development. The agile approach focused on small, frequent changes and software team/component independence. In the world of three-tiered architectures and the related Service-Oriented Architectures (SOA) of the 2000s, agile team structures and the related microservice architectures, focused on creating smaller, independent, and fully deployable software components. Agile and microservices were two sides of the same coin in that they were an attempt to create software that is easier and cheaper to maintain and faster to deploy.

When microservices were widely introduced in the early 2010s, my first impression as an SOA architect was to think "these microservices are just smaller services." That, however, was missing the point. SOA was an attempt to logically break up business functionality into smaller discrete units of functionality. These were often hosted on monolithic hub and spoke architectures such as Enterprise Service Bus (ESB) platforms. This was a good evolution away from the complicated spaghetti code that was often involved in systems that limited themselves to only three tiers and thus three very large applications. The SOA approach was a logical deconstruction of these monoliths into smaller logical components.

The enterprise service approach worked well for breaking apart the logical components of a system, but there was a large amount of interdependence in the applications that used these systems, and hosting became a bottleneck, both in terms of the programming languages used and the underlying compute environments.

By 2014, containerization had made it possible to run an entire server environment on your laptop and deploy your application code to it without needing to wait for system admins or enterprise service teams to hand-deploy your code to centralized servers. With the advent of cloud computing, small teams could write and test their code locally and then deploy their "server" right on cloud infrastructure without waiting for specialized enterprise server teams.

An old joke that encapsulates the idea of containerized microservices involves a developer in the midst of debugging an application error who says to his manager, "I don't know, it works on my machine." His manager responds pithily, "Great! Let's deploy your machine to the server!" And with that, the story goes, the advent of containerized services was born.

One of the much touted features of microservices was that they could be written in any programming language because they were developed and deployed independently from each other. This was another move toward democratizing software development. Now, people who wanted to write software did not need to learn about server infrastructure because it was commoditized in containers, and they did not have to learn the chosen language of their specific company because they could write and deploy code in the language of their choice. Additionally, companies did not have to worry about recruiting specialist programmers who were experts in one chosen language but could focus on hiring experienced developers who not necessarily specialized in a single technology, but who understood the principles of good software design such as Object-Oriented Programming, conditional logic, database design, JSON, and several other programming idioms, and integration technologies.

At this point in the story, if we are tracking with general industry trends, we are somewhere in the 2018–2019 period. The thrust of enterprise software has been broadening access to software development by eliminating the need for highly specialized technical knowledge.

Let's recap. First, distributed computing allowed people without access to a single monolithic compute machine to learn to build and deploy software on their personal computers.

Second, service architecture allowed software developers to write application logic in multiple different languages, breaking down the skill barrier of specialized enterprise developers and allowing developers to mix and match programming languages based on skill level and functional need. That is, basic applications such as data access APIs do not need to adopt all the baggage of the incumbent enterprise languages such as Enterprises Java Beans and C++.

Next, containerization and cloud computing allowed developers and systems engineers to greatly reduce their need for deep understanding of server compute technology. Containers allowed developers to deploy their code in a small, standardized virtual linux server, eliminating the need for constant server maintenance. Enterprise system admins could also simplify their workflows by offloading much of their server compute maintenance to Infrastructure as a Service cloud providers. Small companies were now able to run large enterprise class workloads on demand without the need to employ full-time system admins or acquire expensive hardware.

In 2020, the Covid pandemic accelerated the move toward remote work. The technology was still in its adolescence at the time, and the lockdowns ensured the business culture change that was needed to formalize the new normal of remote work for engineers. Cloud engineering had already moved many of the compute resources to remote locations, microservices had allowed small independent components to be built without massive collaboration overheard, and agile workstyles allowed for easy management of work at a task level. Moving to remote work was not only possible for engineering staff, but it actually made most enterprise development for companies already employing these technologies cheaper and allowed engineers to be more productive. In addition, companies now had access to a much broader recruiting base of engineers since they were no longer constrained by physical geography.

Conway's Law and Remote Work

If you have spent any time around enterprise software architects, you are likely familiar with Conway's Law. Conway's Law is not a real law of physics, but like Moore's Law it describes a tendency in technological development that can be observed and measured in the real world to such an extent that people accept it as a general principle. In its popular usage, Conway's Law represents the tendency of any technical system to be roughly structured in such a way that it has its parallel in the organizational structure of the people who developed it. That means that if your development staff is broken up into three teams–the front end, business services, and database teams, there is a very high likelihood that the components of your systems will follow that same component design, with strong integrations within each of the divisions, but more formalized and complicated integrations between them. The database team, for instance, works with each other every day and thus optimizes all of their components to work together by following the same design patterns and may even be written in the same programming language. There is a high trust between the team members and the components that they build. On the other hand, though, the database team and the business services team in this scenario likely do not work as closely, and so the trust between the team members and the representative components of each of their subsystems includes more formalized integration points, such as formal meetings and standardized API interfaces.

Agile practices allowed development teams to shrink in size and focus on frequent, high-trust interactions between small, focused development teams. With the advent of remote work, we have seen this pattern continue to get stronger as small teams work closer together with greater levels of trust, while extended teams with formalized integrations have become more formalized and distant.

There is a new pattern that is emerging now with the broad-scale adoption of remote work coupled with greatly increased AI automation capabilities. The new pattern is due to a combination of greatly increased developer productivity and remote teams. Technical projects used to take full enterprise teams months to accomplish, but agile teams could perform in weeks. With the productivity introduced by remote work, tasks that used to take agile teams weeks to accomplish can now be performed by skilled engineers in a matter of days. Since the broad availability of large language models such as Anthropic's Claude and OpenAI's ChatGPT, engineers with a general understanding of the problem domain can accomplish many of their weekly tasks in a matter of hours.

We will return to this conversation about the effect on the enterprise workforce later in this book. For the current technical discussion, it is important to understand that the newly automated systems tend to produce their own side effects after some time. The biggest risk is that as these automated systems grow in size, their dependency graph, that is, all the little processes that depend on one another to function can grow to unwieldy proportions, and you can find yourself in the situation described in the following middle-age proverb:

> For want of a nail the shoe was lost.
>
> For want of a shoe the horse was lost.
>
> For want of a horse the rider was lost.

For want of a rider the battle was lost.

For want of a battle the kingdom was lost.

And all for the want of a horseshoe nail.

And while the technology itself such as the programming language or operating system is not the difficult factor, the complications imposed by the developer's process itself become the knowledge wall that must be scaled (to take this battle analogy far beyond its reasonable usage!). I have seen many companies choose to purchase several platform technologies over investing in only one, even at great expense, citing their abhorrence to "vendor lock-in," only to find that they are eventually "locked in" to their own overcomplicated integration processes they built to manage several vendors. We see this same pattern, albeit much more insidious, in the case of neural networks. The logic of the neural networks becomes so complicated that they are regarded as "black boxes," and implementing any one of those systems into core production systems means that you are at its mercy because of the difficulty of reproducing its results independently in another system.

To avoid these issues of overspecialization and "black box" systems, one needs a structured and robust approach to creating applications that are based on high levels of complex automation. These AI applications should be built following the well-established patterns of component design (agents) and explicitly account for the appropriate layers of enterprise quality-of-service features such as testability, observability, maintainability, etc. The rest of this chapter defines a framework from which to think about the overall structure of AI applications, the component design of AI agents, and the AI application stack for organizing your AI applications in such a way that is robust, flexible, and maintainable over time.

Business Process

AI applications are really more about autonomous business processes than specific application functionality and take place at a higher level of abstraction. This means that AI apps can sometimes be encapsulated primarily in a single tool, such as Service Now. Usually to tie these processes into a full application, there needs to be several applications stitched together to provide the end-to-end AI application. In this way, it can be hard to distinguish AI apps from their individual applications. Thus, many companies try to sell a single AI tool, which usually only covers one section of the full application. Dynatrace, for instance, can provide much of the reasoning and store the resource directory, but it must be tied into apps such as Service Now to manage change requests or events that are proactive, not just reactive and based on failure scenarios.

Think of AI applications as working at the same level as an employee. For instance, an employee who has the responsibility of managing accounts receivable for a company is not usually hired solely due their expertise on a single accounting system, such as Quickbooks, nor is their job description limited to utilizing only that tool. The value of the accounts-receivable employee is the fact that they understand and can manage the business processes that are included within the accounts-receivable capability of a business. This would include the management of processes related to financial accounting, bank integration, customer invoicing, debt and payment collection, payment processing, customer and management reporting, and financial forecasting. For a smaller company, a single accounts-receivable employee may be responsible for much more than this, including certain operations and vendor management functions.

Employees may be hired based on their experience with certain tools or software applications, but it is a very poor accounts-receivable employee who refuses to work with any tools outside of Quickbooks. Any number of other applications are used in their day-to-day job duties such as third-party websites, home-grown applications, pdf documents, and of course word-processing and spreadsheet software. In fact, the lingua franca of any finance-based role is the spreadsheet, and this is typically in the form of Google Docs or Microsoft Excel. As much as software vendors like Intuit would love nothing more than for everyone who engages with the accounts-receivable duties to only work in the Quickbooks tool, the reality is that business capabilities will always encapsulate a diverse set of tools.

While business capabilities require many tools to function, it is likely that a particular business focuses their capability around a single tool, such as Quickbooks. When defining your AI applications, though, one must be careful not to identify the AI application with the specific functional tooling that it utilizes, in the same way that it is not a good practice to hire an employee with the expectation that they only work with a single tool. When most vendors from specialized software solutions offer us their custom AI tools, this is the equivalent of hiring that "Quickbooks Guy," for instance, but unfortunately these AI tools are not as adaptable as humans, and vendors are not apt to sell you a solution that will easily let you move from their own solution to a competitor's solution, so purchasing the Quickbooks AI will not provide you with the high level of flexibility and intelligence of a fully matured AI application.

Defining Your Business Ontology

If you took a college course in philosophy, you probably came across the branch of philosophy called ontology, which is traditionally defined as the study of "the nature of being." That's all well and good for a weekend of reading the *Meditations of Decartes*, but in business and computer systems engineering, the term ontology is specified as *business ontology* or *AI ontology*. Regarding Enterprise AI applications, it's a mix of both of these more pragmatic uses of the term.

Business ontology refers to the concepts and terms we use to describe our business. It refers to what our business does, how it does it, who is involved in doing it, and why it is done that way. Business strategists for decades have been defining business ontologies to help focus a vision and purpose for the business. For instance, consider McDonald's. McDonald's as a corporate entity has two main sources of revenue. One is to manage the brand, equipment, and operating procedures as a franchiser. That is, they are not interested in running the day-to-day operations of local restaurants–they (mostly) leave that to franchisees who buy into their business. The second form of revenue for McDonald's is through real estate speculation. They purchase prime pieces of real estate and lease it to the franchisees who run their business. Their real estate speculation is often so good that a purchase of land by McDonald's is an early sign indicator of rising real estate prices in the area.

The point of this discussion about McDonald's is that it illustrates an important point about understanding the nature and purpose of your business. When one realizes what McDonald's is really about when it comes to revenue streams, one can start to understand the reasoning behind their business decisions.

At a lower level, business process analysts have also dealt with ontologies for a very long time, though they would not describe them this way. Lean and Six Sigma are frameworks for identifying the underlying activities of a company for the purpose of optimization and cost savings. An understanding of the strategic Why of the business is a crucial guide for these practices, but an ontology that only includes the Why and doesn't bother about the What, How, or Who is not very useful, and in fact is no more than a hopeful concept. The business process analysis portion of business ontology is what gives a business its skin and bones.

The term AI ontology has become popularized in the past few years and defines the conceptual domain of a trained AI. For instance, an AI that is specialized to work in a warehouse to move boxes has a very small ontological domain. There is no reason for the warehouse robot to know anything about cartesian philosophy, for instance. OpenAI's ChatGPT models, on the other hand, are created with the express purpose of pursuing General Intelligence, that is, they are meant to understand everything in the human experience. For ChatGPT, it has no problem writing a thesis on Descartes Meditations, since that is exactly what it was trained to do. Ironically, this general knowledge is not so useful in our warehouse situation and for this reason, we stress limiting agents to a specific sphere of operation and then combining them for more complex tasks.

Business + AI Ontologies

One of the interesting features of remote work is that it has greatly accelerated the move to digitization for almost all companies. This not only means that we don't use nearly as much paper as we used to by printing out and distributing reports, we also have started using tools that focus on digital collaboration among whole teams. For instance, when we needed access to a particular project report, it was easy to find our project manager's cubicle and ask them to email you the latest report. Since remote work is primarily asynchronous in nature, we have now begun to rely more strongly on small teams rather than individual interactions. Part of the reason for this is that the "face-to-face" of one-on-one communication is not what it used to be. More significantly, workers are beginning to optimize their time to a much greater extent and will now request that artifacts such as project reports are stored in shared locations in Teams, Sharepoint, Confluence, Slack, etc. Remote workers have normalized the use of these tools out of necessity, and in response these collaboration tools have gotten much more reliable and full-featured than they were before the pandemic.

With business documentation moved toward greater centralization and away from point-to-point private communication such as email and text, it is now possible to reliably construct full business ontologies from shared online corporate sources.

We can now create AI agents who can take on the role of particular business actors and consume all of the ontological artifacts that define what our business looks like. With these business agents in place, we can begin to work with these AI business personas as if they were actual team members by asking them questions about the business, requesting analysis of business processes, and performing business tasks.

General AI vs. Business AI

Several years ago I worked on a study concerning AI and business ethics for the ACM (Association for Computing Machinery). One of the stickiest concerns was dealing with the idea of virtue. In the case where multiple virtues (honesty, self-control, generosity, etc.) are in play with one another, and perhaps even conflict with each other, context plays a vital role. A virtuous decision depends strongly on the situation in which the moral agent finds themselves in. A spy or a captured soldier, for instance, could choose the values of loyalty and self-control as more important when across enemy lines than the value of honesty. In fact, the value of honesty itself is always dependent on other values in order to be virtuous. We do not find people who claim to "just say it like it is" to be particularly virtuous if they are offensive or partake in gossip.

To solve this ethical dilemma for AI, we tied the ethics to a very specific list of codified business guidelines that were limited to the specific business context in which the AI operated. While it may be difficult to define the general ethical principle of "freedom of speech," within the context of business ethics, the problem can be narrowed and specified to a list of rules such as "avoid harmful speech," "do not divulge company intellectual property," "do not divulge privileged financial information to unauthorized groups," etc.

Similarly, the right strategic solution for a company will also be highly dependent on the situation a particular company finds itself in. General AI, which has been trained on thousands of business scenarios and public economic data will be inadequate to make real-time strategic recommendations for a company without understanding the particulars of its operating model and business goals. A general AI solution can go so far as to tell you "companies that look like yours have made these kind of decisions in the past." That can be useful information, but there are two key limitations of that statement. The biggest issue is that as a business you are likely trying to differentiate yourself from the competition. Your differentiation relies on your unique vision and goals for your company. Just like in Reinforcement Learning (which we will address later in this book), these goals must be fed to the AI in order for it to optimize its responses to meet those goals. The second issue is the phrase "companies that look like yours." This is a matter of distance and perspective. At what level of abstraction does a company "look" like yours. If I run a local hamburger stand, does my business look more like other small restaurants that are my size, or do I look more like McDonald's. This depends on which perspective you are taking, and it also depends on the level of detail the AI understands your business. Your corporate ontology is the lens through which an AI application understands what your company looks like.

Components and Events

A corporate ontology is made up of two broad categories–components and events. Simply stated, a component is a noun, and an event is a verb. A component does something to another component or gets acted on by another component. Traditionally, we have focused on capturing the components of our business, and this is in large part because it is easier to conceptualize and track those things. We pay employees, and we have a database of all of our employees. We pay employees and use a physical building. We purchase computers and printers and large enterprise software licenses. Often we have invoices and databases to track these components, so it is fairly easy to keep a digital record of them. What is more difficult is to keep record of all of the interactions between these components in a standardized and consumable fashion. An employee updates an invoice. Our accounting software updates our ERP system. Customers book consultations with our employees or purchase our products online. These transactions are collected, but they exist in independent systems that are isolated from one another.

Companies with mature IT organizations have typically made hefty investments in data warehouses that centralize the collection of these components and events across the company. While these efforts are useful for historical reporting, they require upfront work in the normalization and canonicalization of these business interactions across the company. These massive standardization methods are the reason why roughly half or more of large enterprise data warehouse initiatives go over budget or fail to deliver on their promises.

The promise of AI in these cases is that once a company is able to define its ontology (its goals, events, and components) at a high level, the process of creating standardized canonicals and schema formats is largely taken care of by the LLMs themselves, since they can understand documentation in place, whether the information resides in data logs or wiki documents. Enterprise data warehouses, data lakes, and other such technologies rely heavily on data manipulation through the use of standardized data schemas, and then the interpretation of the data is performed through static reports read by human analysts. AI applications utilizing LLMs can bypass the data wrangling stage altogether and can spend more time analyzing the business process events as they happen throughout the day.

Figure 4-1 shows an example of a strategic Enterprise AI application that is built on a composition of several departmental AI applications. This is similar in design to the enterprise data warehouse, but the abstraction is at a much higher conceptual level, removing much of the existing complexities around technical schema translation layers.

61

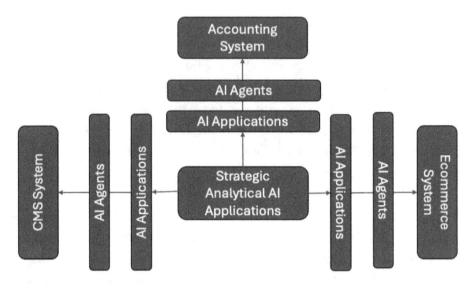

Figure 4-1. *A Strategic Enterprise AI Application Integrating Departmental AI Solutions*

AI Agents and AI Applications

An AI application is made up of a combination of limited-function AI agents. AI agents are composed together to make up a single AI application. AI applications, similarly, can be composed to make higher level AI applications. This is the common object composition pattern that has been fundamental to technology service architectures throughout the last couple of decades.

There are several vendors on the market who are happy to sell enterprises expensive AI solutions. While these solutions are not without merit, it is important to understand the differences in third-party AI services that are offered on the marketplace and the kind of Enterprise AI applications which are described at length in this book.

Platform as a Service, Vendors, and AI Applications

Platform as a Service providers such as Microsoft Azure, SAP, Oracle, and others offer businesses their version of a single technology solution to run their full business. By paying for these turnkey solutions, businesses are promised a single technical platform that is seamlessly integrated among all of the capabilities of your business. Most vendors offer the same vision of a technical platform but within a single business capability. Intuit, for instance, provides a solution that will perform all of the business functions within your accounting and finance areas. For large companies, Dynatracc will provide a single set of observability tools that will monitor all of your business processes and technical systems within one tool.

While it is possible to purchase one of these large platforms and rely on them to provide your AI applications, the limitations inherent in this approach are similar to the current issues of "vendor lock-in" that businesses face today. Microsoft, in particular, has done a very good job of creating a full ecosystem of enterprise-scale tools within which companies attempt to make exclusive investments. For instance, many companies will refer to themselves as a "Microsoft Shop," meaning that they have focused their technology investments on the Microsoft ecosystem in order to make integrations easier within their various technical tools and to manage their software acquisition expenses through a single vendor–that is, Microsoft. Other Platform as a Service providers have also done this, such as SAP and Oracle, but Microsoft is generally the product most are familiar with. I leave AWS out of this scenario because while they are a technical Platform as a Service company in competition with Microsoft, most of their efforts are not positioned toward Platform but rather infrastructure. AWS provides a vast array of business capabilities and technical services, but these are all focused on winning over your company's compute spend. That is to say, AWS wants to be your company's exclusive data center. This approach has its merits, but in terms of the AI applications we are speaking of here, AWS does not itself provide a comprehensive enterprise AI application solution set, as the other Platform as a Service providers do. While Google provides some excellent marketing solutions, they have typically focused their enterprise efforts in the Infrastructure as a Service capabilities along with AWS.

The Downsides to Single-Vendor AI Application Solutions

There are two downsides to choosing a single vendor to provide your AI application solutions. The first is cost. Large vendors such as SAP provide a feature rich set of enterprise capabilities but are usually priced so that only very large companies can take advantage of their services. For companies that are large enough to be traded in the US Fortune 500, their services can be a solid value proposition because the size and complexity of their business operations warrant such an expense. For small-to-medium-sized companies, however, these solutions that run into the several millions of dollars per year are not financially viable.

The second downside to single-sourced AI applications is that they lack the flexibility and customization that are often needed within a dynamic business environment, particularly a small business that is growing and evolving. Platform as a Service vendors operate on the assumption that all tooling for your business can be sourced from a single platform vendor. While this may be true when it comes to large enterprise applications such as relational database systems, directory services (such as LDAP), and enterprise resource planning (ERP), the tools that are closer to the end-user are not centrally controlled by a single corporate IT department and are much more dynamic in nature. For instance, a single accounting department likely works with upward of 10 or 15 tools as it integrates with vendors, external customers, and various departments within the corporation. An example of just some of the various end-user tools used by an accounting department on a given day may look like this:

- Specific web APIs

- Spreadsheets

- End-user databases such as Access

- Accounting software

- Word-processing software

- Productivity software such as clipping and screenshot software specific to Apple, Windows, etc.

- Videoconferencing and chat software

As they deal with other individuals outside of their corporation, they could be dealing with two or more vended brands of each of these tools. Traditional platform-based software works deep into the fabric of corporate IT and is technical by its very nature. The platform software requires experts to implement and maintain it, and so access is understandably restricted and centralized to make sure it is managed appropriately by engineers who have been specifically trained in that particular technology tool. Because AI applications are concerned with business functions and business capabilities, the tool integration is only a specific piece of the larger AI application.

Large language models and general-purpose AI agents by their nature operate at the end-user level and are able to understand and integrate these various tools with ease. AI agents provided by the individual vendors for these tools can make this integration much easier, and is indeed a selling point for them, as the multiplicity of tools and integrations within a business process is not ideal whether the business process is handled by a human or an AI.

AI Agents

The AI agent is the building block of AI applications, so we will start with agents and work our way up the stack from specialized tool-based agents up to complex, higher level AI applications.

AI Agent Hierarchy

AI agents are divided into the following four types, moving from specific functionality, which is tool integration, to the highest level of complexity with Business Logic Sets, as shown in Figure 4-2:

- Tool integration
- Information set
- Business function
- Business logic set

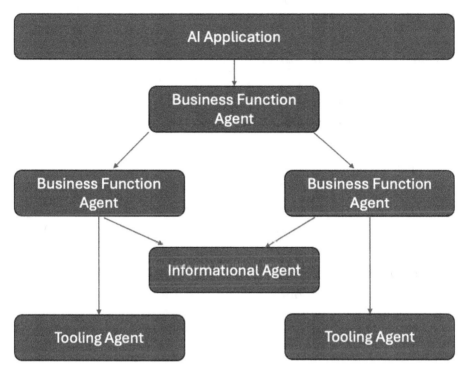

Figure 4-2. *An example of agent hierarchy*

AI agents are often provided by specific vendor solutions, such as a Quickbooks Invoicing Agent, Salesforce CRM agent, Service Now ITIL agent, and so on. Some large business platform vendors such as SAP and Microsoft offer agents at the business process level within their product, but these come with the limitation of being tied to their own vended ecosystem.

Tool Integration Agents

Tool integration agents are exactly what they sound like–they are AI agents that allow you to easily integrate with a specific tool. The interface could be a conversational chat that takes place either with other agents or other humans. This would be a way of dynamically integrating with the tool in real time. Another integration pattern may be through the use of a conversational development tool that is used by a human handler or developer to set up formal integrations or rules (such as security or other guardrails) within the tool to set up and define business processes.

These integration agents will almost always be provided by the tool vendor, though it is possible that specialized integrations such as through the use of a collection of web APIs could be developed and maintained internally, though these homegrown integrations will be fragile and likely short-lived, replaced eventually by the vendor's own integration agents, or by moving to new tools that provide their own integration agents.

Information Set Agents

AI agents work best when they act as experts in a particular topic. Large language models (LLMs) in particular are very good at understanding small sets of information in the form of technical documentation and conversational chats. Ironically, LLMs in a business setting are much more effective using small rather than large amounts of data. Big data is excellent for training the language models, but once trained, LLMs work best against small, well-curated information sets. Curation of information becomes increasingly important as access to various types of information becomes more immediately accessible than ever before.

Business Function Agents

Business function agents work at the level of particular business functions. These granular functions include such tasks as:

- Sending a weekly email from a CRM

- Executing weekly payroll tasks

- Performing security access request workflows

- Distributing monthly financial reports

- Managing external information feeds

- Communicating with external buyers

- Publishing new social media content

When placed in a high-level list, it seems that any number of existing automation tools could perform the tasks listed above. As we will see our Whaling Airport fictional examples, these business function agents go far beyond basic operational automation.

Much of my practical experience with AI has been in the area of AI Operations, and before the days of wide-scale AI, I spent a lot of time in the 2010s creating automated systems. I once managed a team of systems admins who spent much of their time writing scripts in languages such as puppet, chef, and powershell. When I stepped into the role of manager for this department, much of the department was overworked and would spend long nights running maintenance jobs. Over the course of a couple of months, the situation seemed to get worse, even though the team was telling me that they were spending as much time as they could writing automation scripts to run these tasks.

After some time, I discovered that this team was indeed writing scripts for all of their operations, but the scripts were almost completely procedural with no decision points. For instance, a script would look like this:

1. DO X

2. DO Y

3. DO Z

The system admins would log in at 11 at night and run this job if certain conditions from the previous business day were met. Or they would have batch jobs set up to run these, but fail out and alert the team if there was something amiss, so they could figure out what to do next.

The difference between procedural scripting and business automation comes down to automating the decision-making. In the situation I have been describing, the factors that needed to be taken into consideration were too numerous, according to the admins, to trust automation code to make the right decisions.

Business Logic Set Agents

Business logic sets are compositions of several business process agents. Monthly financial reporting, for instance, should consist of several specific business functions. When new vendors are added, or new input streams, these should come in the form of new business logic, informational, or vendor-specific agents.

A Word on Abstraction

I have identified four different levels of agents, each at increasing levels of abstraction. If you have a background in software development, you will notice that this follows a similar pattern of Object-Oriented Design and Service-Oriented Architecture. Decomposing logical functionality into different levels of abstraction is a well-known approach to designing software and has many benefits. These approaches also come with a major downside, which is overcomplexity. While very large projects may take advantage of each of these levels of abstraction in their agent design, most small companies or independent consultants should avoid overcomplicating their solutions with too many layers of abstraction.

It has been my experience that erroring on the side of simplicity in software design is usually the best approach for maintainable systems. One does not want to tie all functionality to a small set of interconnected tool integration agents without any level of abstraction, as this could make your solution overly rigid and unable to evolve over time as new techniques and tools become available. On the opposite extreme, creating a layer of idealized but ultimately useless abstraction brings additional complexity to all of your components and inflates the price of development and long-term maintenance beyond reasonable levels. As you can imagine, small companies and startups usually opt for the first approach and are eventually faced with the unfortunate choice of a painful refactor or a complete rewrite of their systems in order to scale to meet growing business demand. Large companies with large internal staff tend to move toward the other extreme, which seems inevitable as the nature of a large IT staff is to create large, complex systems.

Make no mistake, automation is difficult. The first step, however, is to document the business functions that need to be automated. In most cases, undocumented business processes that are tribal knowledge seem much more complex and nuanced than they usually are. AI agents can be trained to follow precise instruction sets that look less like AI and code and more like a standard process and procedure checklist.

The difficulty usually lies in the ability to easily modify these instructions, audit the historical runs of these automated functions, and provide the proper observability and alerting functionality needed to confidently run these agents completely independent of direct human supervision. That is where modern AI agent tools, coupled with the AI application stack architecture, come in.

AI Application Stack

Large enterprises that take advantage of SaaS platforms will focus on defining their business processes within a set of tools to define their AI apps but will typically be end-users of several of these tools and not build out every piece. For instance, a retail company will typically not spend the resources required to train its own LLM from scratch, which not only requires a team of specialized data engineers but also can cost over $100,000 in cloud compute resources and several months to train. The approach most companies who are end-users of these tools will take will be to configure each of these pretrained AI models and perform targeted training based on their own company data and ontological framework. There will be coding involved, or at least detailed configuration, but this will be focused on building specific business process steps into a stand-alone AI agent, integrating specific data feeds, validators, and reporting features.

By utilizing a stacked architecture, AIOps developers can encapsulate separate features of the AI app into distinct areas of concern. This will act as a way to understand specific capabilities of turnkey operations' applications such as Service Now or Dynatrace and break down their functionality into discrete capability areas that can be evaluated on their own merits. For instance, Dynatrace Davis AI performs excellent reasoning facilities for server and application operation performance, but Dynatrace does not offer capabilities to predict or perform actions based on user-inputted requests.

This architecture stack also serves the purpose of a checklist. Validation and auditing, for instance, are critical pieces that most tools with AI reasoners only partially provide or do not provide at all. For full-cycle AI apps, it is also important to allow for tools for human intervention at critical stages along the process as part of the configuration capability. Custom AI apps will often stitch together an automated system that works well in 90 percent of circumstances, but 10 percent of the time, it must be stopped and manual intervention or bypassing all together has to take place. Taking the time to build configuration into the system can increase the robustness of the app and increase the likelihood of long-term adoption and evolution as the business processes morph to meet changing needs.

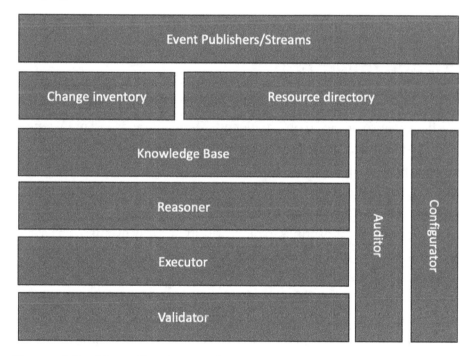

Figure 4-3. *AI application stack*

Event Publishers/Streams

From a business workflow perspective, AI applications start with an event stream. Typically, the application will subscribe to a published event source that indicates that some significant action has happened and needs to be acted upon. The state of the art today in terms of streaming event publishers is Apache Kafka, but there are other publishing technologies that can be used. Enterprise Service Buses such as Oracle Service Bus, or hybrid stream/batch platforms such as TIBCO could also act as a subscription source. Some events may be directly based on dynamic user or system actions, and others may be based on temporal factors, such as scheduled events such as financial reporting schedules or maintenance schedules.

Change Inventory

Changes to the system are typically predefined as a set of agent instruction sets but can also be based on broader goal-based directives. For instance, many AI-based security tools will regularly scan an enterprise environment for open ports or misconfigured servers for security vulnerabilities. A developer does not provide specific instructions in this case, rather the vendor of the Security Ops tool provides general direction for "securing an enterprise" and subscribes to high-priority security events such as the publication of new vulnerabilities and security patches, to which the security AI tool responds.

Once these changes are recognized, they are added to a library of changes to be scheduled and performed. In the case of high-priority security events, these events are performed in real time. In other cases, the change items may be logged and passed on to human handlers for approval before moving to the execution phase

Resource Directory

The resource directory is a record of all of your component assets that are involved in business events. The simplest way to think about this from an IT perspective is the CMDB (Change Management Database) found in tools such as Service Now. As specific changes are identified in the change log in response to business activities consumed in the event streams, the change log provides a list of candidate components that these changes can be performed against.

Resource directories are difficult to maintain as they are often manually created at a point of time and then go out of date within a period of a few months. There are many tools that can scan an enterprise for new Infrastructure components. With AI tools, it is possible to scan an enterprise for activity streams that take place at the business process level and identify components and events that correlate together as discrete business functions. These functions can be classified as independent components and placed into a dynamic resource directory. Vendors can often do this today within their own systems, but Enterprise AI applications will be able to act independently from specific vendor tooling and identify these higher level cross-functional processes within the application, giving greater insight and optimization opportunities to business leaders and AI handlers.

Knowledge Base

As we have seen, general AI intelligence is useful for certain very high-level tasks, but for specific business functionality, especially where process-level decisions are being made, the knowledge base from which a particular business agent is operating should be as narrowly focused as possible. While language models are good at decrypting vast amounts of information, it is likely that the larger the knowledge base of operating procedures fed to an agent, the more likely there will be conflicting sources of information. Different departments may have slightly different approaches to completing similar workflows. While it is useful to have AI applications that can perform analytics across an enterprise to find opportunities for optimization, AI agents that are focused on performing low-level business tasks should be focused on specific information sets from which to operate. These knowledge bases often come in the form of RAG (Retrievel Augmented Generation) documentation such as wikis, pdf libraries, code bases, databases, and other forms of documentation.

One of the main advantages of this specialized approach is for maintaining and debugging large AI applications. When a separation of duties between agents and their respective knowledge bases is maintained, it is easier to pinpoint where certain anomalies in instruction sets are located and minimize the blast radius of erroneous documentation or even maliciously introduced instruction sets.

Reasoner (What Should Be Done)

While the change log identifies a set of potential task level changes, the reasoner provides awareness of the business context at a higher level of abstraction that can more appropriately make decisions about what tasks should be performed, how they should be ordered or scheduled, and what should happen when particular tasks fail or perform unexpectedly. For many, this is the "AI" part of the AI applications, as much of the duties so far can be provided by existing enterprise tools and business rules engines utilizing essentially procedural logic.

Much of the processes in the reasoner component rely on processes such as anomaly detection at the process and information levels. Human feedback can also be incorporated at this stage to provide model corrections and reinforcement goals.

Executor (Do the Change)

In the continued spirit of separation of duties, the executor is separated from the reasoner. The executor may consist of several small agents who have specialized security access to critical systems. Certain actions may also require human approval before execution is performed, so another layer of human feedback could be introduced for certain privileged actions.

Validator (Did It Perform the Right Change)

Every change should be followed by a set of automated validation steps. The first level of validation is a simple health check. These health checks should be run periodically throughout the day to ensure the continued reliability of the system. The second level is basic application-level checks. These are validations at the business process level and are more involved than basic health checks, which may only check that a login page or information set is available for a given application. Business process level validations can get complicated, such as validating the ability to perform financial transactions via credit card. For this reason, these tests are often harder to design and take longer to run, and therefore run on a less periodic basis than basic health checks. AI tools that are aware of the underlying business processes will be much more capable of discovering and creating candidate business process validation tests.

Auditor (Reporting)

As AI can react and adapt to events at a much faster rate than humans, it is important to have an audit trail available for ongoing analysis and historical audit purposes. The auditor component plays an important role in the regular handling of AI applications. Most platforms do not provide the level of auditing and reporting that is appropriate for this step, so this will almost always have to be built in house.

Configurator

Large projects focused on process automation often start off to great excitement and fanfare, but they require ongoing maintenance and updates as processes evolve and change. The context of any business is dynamic and subject to continual change, both externally in the form of economic factors and internally in the form of staffing changes, but regress back into semimanual processes that can sometimes become even more complicated than the original process since more systems may be involved than before. The configuration component of AI applications is critical to making sure the application is maintainable, long-lasting, and doesn't end up creating more complicated processes over time than the original processes it was designed to automate.

The first key to the configurator component is the ability to make changes to any of the decision-making logic within the AI application with as little effort as possible. It is important to plan for this kind of functionality early on in the design of your AI application, as it is much more difficult to retrofit configuration hooks into an existing system. Adding on configuration points of entry will be inevitable at some point, as it is impossible to anticipate all of the ways your application will perform in the real world, and tweaks will be needed, but building your components with this functionality in mind is key to a successful application.

The second piece of functionality that is included in the configuration component is the ability to roll back changes based on events that recognize the auditor component. This means that one must keep a version history of states of the application to roll back to. If an undesired state is left running for too long, the system morphs to accommodate the change (many times in undesired ways), and it makes it much more difficult to unwind. Finding and stopping misconfigurations or bad outcomes as early as possible is what this configuration component is for, and there should be a small number of "break the glass" configurations in place that will stop the application in part or whole until the problem is resolved.

Conclusion

In this chapter, we have talked about the evolution of enterprise applications from Service-Oriented Architecture to microservices to Enterprise AI applications. Business processes are the key to AI functionality, and these processes are laid out logically in business ontology.

This chapter focused on the high-level design patterns of AI agents and AI applications that inform an enterprise architecture. In the next few chapters, we will dive deep into the specific technical features of AI applications that are utilized within the components that make up the Enterprise AI application stack.

CHAPTER 5

Data, Machine Learning, and Reasoners

Introduction

In the realm of artificial intelligence, data and machine learning algorithms form the bedrock upon which intelligent systems are built. Data provides the raw material that fuels machine learning models, enabling them to perceive patterns and make predictions. Meanwhile, sophisticated algorithms transform this data into actionable insights, driving decision-making and automation. This chapter delves into the intricate interplay between data and machine learning, exploring their respective roles, challenges, and best practices for their effective integration within AI systems.

Artificial intelligence encompasses a vast landscape of technologies aimed at building intelligent machines that can perceive, learn, reason, and assist humans in making decisions and accomplishing tasks. At the core of many AI systems are data, machine learning algorithms, and logical reasoners. Data provides the raw material for teaching AI systems about the world. Everything from images, video, audio, text, and sensor readings

A. Cagle and A. M. C. Ahmed, *Architecting Enterprise AI Applications*, https://doi.org/10.1007/979-8-8688-0902-6_5

can train AI models to recognize patterns and make predictions. Massive datasets along with increased computing power have fueled the recent AI boom, enabling more complex neural networks to achieve state-of-the-art results on tasks like computer vision, speech recognition, and natural language processing. However, data alone is not enough. Machine learning provides the algorithms that actually learn from data and create predictive models. Popular techniques like deep learning use neural networks with multiple layers to learn hierarchical representations of data. Other approaches include decision trees, support vector machines, Bayesian networks, reinforcement learning, and more. Choosing the right machine learning approach involves weighing factors like accuracy, interpretability, computational complexity, and the type of available data.

In addition to learning from data, AI systems also need the capacity for logical reasoning in order to make sound inferences. Knowledge representation and reasoning techniques equip AI agents with common sense facts and rules that allow them to deduce new information. These reasoners complement data-driven learning, helping AI agents generalize knowledge to novel situations. Large language models that can understand and generate human language have recently emerged as a third critical component alongside data and reasoners. Systems like GPT-4-o and Claude 3.5 Sonnet exhibit an ability to reason about text passages, answer questions, and follow instructions. While lacking robust logical reasoning, their sheer scale and exposure to massive text corpora allow them to mimic certain reasoning capabilities. Bringing together data, machine learning, reasoners, and language models enables the development of intelligent agents that can perceive, learn, reason, and interact naturally with humans. However, effectively integrating these components involves overcoming challenges like data bias, transparency, and aligning AI goals with human values.

This chapter provides an overview of the capabilities and limitations of current techniques across all these areas. We discuss best practices for applying AI responsibly to solve real-world problems. Readers will gain key insights into how to leverage the strengths of data-driven learning, logical reasoners, and language models while mitigating risks toward building trustworthy AI systems.

Data: The Backbone of Machine Learning

Data Types and Formats

Data powers machine learning. Whether structured or unstructured, clean or messy, data provides the critical raw material for training machine learning models. As the adage goes: garbage in, garbage out. Without good data, even the most advanced algorithms will fail. This chapter provides an overview of the key data concepts and preprocessing techniques essential to building effective machine learning systems. We explore the various data types and formats, highlight common data quality issues, and discuss critical steps like cleaning, normalization, and feature engineering. By understanding data's indispensable role and learning how to wrangle it into shape, we lay the groundwork for extracting powerful insights with machine learning.

Structured data lives in neat rows and columns within organized databases or spreadsheets. This includes numeric data like temperatures, dates like calendar events, and categorical data like product names. Structured data is easy for both humans and computer systems to interpret and analyze thanks to its tidy organization. Common structured data sources are SQL databases and Excel files. For example, a retailer might track each product sale in a SQL table with columns for date, product ID, quantity, and price. This structure enables convenient filtering, aggregation, and analysis. Structured data is foundational to many business operations and analytical processes. Its clear format not only aids in straightforward querying and reporting but also serves as a reliable source for more complex data analysis techniques like predictive modeling and machine learning.

In contrast, unstructured data lacks innate organization and formatting. This includes free-form text like social media posts, images like X-rays, audio like call center recordings, and video like surveillance footage. While rich in information, unstructured data poses challenges for analysis since key details are buried within multimedia formats. Natural language processing (NLP) and computer vision techniques help extract insights from unstructured text, speech, and images. NLP can analyze sentiment in customer reviews, summarize lengthy documents, and even translate languages. Computer vision enables the recognition of objects in images, facial identification, and medical image analysis. Despite these advancements, transforming unstructured data into usable features remains an active area of research. Issues like ambiguity in language, varying quality of audio recordings, and differences in image resolution add complexity to the task. Additionally, the sheer volume of unstructured data generated daily requires robust processing capabilities and sophisticated algorithms to handle efficiently.

Moreover, the integration of unstructured data with structured data is important for comprehensive analysis. For instance, combining text data from customer feedback with sales figures can provide deeper insights into product performance. Effective methods to bridge the gap between structured and unstructured data are continually evolving, aiming to enhance data-driven decision-making across various domains. Unstructured data holds immense potential for uncovering hidden patterns and trends, making it a valuable asset despite its analytical challenges. As technology progresses, new techniques and tools are being developed to better harness the power of unstructured data, driving innovations in fields ranging from health care to marketing.

As the name suggests, semistructured data falls somewhere between structured and unstructured data. While not as neat as database tables, semistructured data retains some organizational properties. Common examples include log files, XML documents, and JSON records. For instance, an ecommerce site log file tracks each user click event with the timestamp, product ID, and session ID. While not a database table, these events adopt a consistent structure. By parsing the logs, key metrics like products viewed and sessions per user emerge. Semistructured data is flexible, allowing for variations in data types and structures within a single dataset. This flexibility is particularly useful for web technologies and applications that require dynamic and scalable data formats. Semistructured data formats enable easier integration and data exchange between disparate systems, facilitating interoperability in complex IT environments. Analyzing semistructured data often involves specialized tools and techniques. For example, parsing JSON or XML documents requires understanding their hierarchical structure, and querying such data might involve languages like XPath or JSONPath. While more challenging to manage than structured data, semistructured data offers a balance between flexibility and organization, making it a valuable asset for many modern applications.

Data Preprocessing

Real-world data often arrives in a messy, incomplete, and sometimes inconsistent state. Before machine learning models can be effectively applied, data must undergo preprocessing to ensure it's in a suitable format. This process, known as data wrangling or data preprocessing, transforms raw data into a clean, structured dataset that enhances the performance of predictive models. Handling missing values is one of the critical tasks in data preprocessing. Techniques such as imputation, where missing values are filled in using statistical methods like mean, median, or machine learning algorithms, help maintain data integrity. Alternatively, data points with missing values may be selectively removed if their impact on the dataset's integrity is minimal.

Data cleaning involves correcting errors, dealing with outliers, and standardizing formats. This step ensures consistency and reliability in the dataset. For example, text data might be normalized by converting all characters to lowercase and removing special characters, while numerical data might be scaled to a uniform range to prevent certain features from dominating others. Feature encoding is essential for transforming categorical variables into numerical formats suitable for machine learning algorithms. Techniques like one-hot encoding, which creates binary columns for each category, or label encoding, which assigns numerical labels to categories, enable algorithms to process categorical data effectively. Feature selection helps reduce dimensionality and computational complexity by identifying the most relevant features for the prediction task. Statistical tests, feature importance from models, or domain knowledge-based approaches are used to select features that contribute most to model performance. Feature transformation involves creating new features from existing ones or transforming features to meet the assumptions of specific models. Techniques like generating polynomial features or applying logarithmic transformations to skewed data can enhance the predictive power of models. Data integration involves merging data from multiple sources to create a unified dataset suitable for analysis. This step ensures that all relevant information is considered, providing a comprehensive view for modeling and decision-making.

Data cleaning is also called data scrubbing, the cleaning process targets imperfections like errors, outliers, and missing values. Techniques like data validation check for outliers and invalid entries, while imputation fills in missing values to complete the data. The goal is to resolve issues upfront rather than allow dirty data to mislead downstream analyses. To enable fair comparison between data points, we often standardize values to a common scale. Normalization rescales the data to fit within a specific range like 0 to 1, while standardization shifts the distribution to have a mean of 0 and standard deviation of 1. Different algorithms benefit from different standardization methods.

Real-world data frequently appears in formats unsuitable for analysis. Data transformation converts raw data like text categories or image pixels into numeric features digestible for machine learning models. For example, encoding techniques convert text labels into numeric categories. Derived variables like Body Mass Index (BMI) transform height and weight into a health metric. Dimensionality reduction through principal component analysis (PCA) distills thousands of interrelated variables down to fewer underlying factors. Feature engineering crafts informative input variables that connect raw data to target outcomes. Domain expertise guides combining, parsing, and transforming data fields into powerful predictive features. As an example, timestamp data can be engineered into time-based indicators like hour of day, day of week, and seasonality that may correlate with patterns in the target data. Feature engineering is both science and art–part analysis, part creative exploration.

Data Quality

"Garbage in, garbage out" underscores the pivotal role of data in machine learning. Issues such as insufficient data volume, inaccurate labels, and biases in sampling can significantly undermine the performance and reliability of machine learning models. Conversely, high-quality data forms the bedrock upon which accurate predictions and impactful insights are built. Establishing rigorous data collection processes is the first line of defense against data quality issues. This includes defining clear data acquisition protocols, ensuring data completeness, accuracy, and consistency at the point of entry. Continuous monitoring and validation frameworks are crucial to maintain data integrity over time, identifying and rectifying discrepancies or anomalies promptly. Cross-validating model outputs against real-world benchmarks is essential to validate the effectiveness of data and model performance. It helps surface cases where poor data quality leads models astray, enabling data scientists to refine algorithms and improve data quality iteratively. This feedback loop not only enhances the accuracy and reliability of predictions but also fosters trust in the insights derived from machine learning models. In the era of exponential data growth and diverse data sources, achieving perfect data is often impractical. Instead, the focus shifts to ensuring "good enough" data quality where the data provides sufficient accuracy and utility to drive meaningful business decisions. Data quality, therefore, becomes less about achieving perfection and more about optimizing for the specific requirements of the analysis and application at hand.

Data serves as the lifeblood of machine learning systems. Understanding the different data types and formats, mastering the preprocessing techniques that transform raw data into informative features, and prioritizing data quality are fundamental to extracting actionable insights. Just as quality ingredients are essential for creating a gourmet meal, high-quality data forms the cornerstone for advanced algorithms to uncover valuable patterns and trends in complex datasets.

Machine Learning Algorithms
Overview of Machine Learning

Machine learning, a branch of artificial intelligence, empowers computers to learn patterns from data, making predictions or decisions without explicit programming. It's like teaching a child to recognize flowers by showing countless examples instead of defining rigid rules.

In 2024, my friend Hala, an attorney starting at a new firm, faced a peculiar challenge where she was working in a specialty she hadn't tackled before. She was trying to streamline her workload, but the case patterns and client behaviors had become increasingly erratic. A simple calendar reminder wasn't cutting it anymore. Enter machine learning. Machine learning, a branch of artificial intelligence, empowers computers to learn patterns from data, making predictions or decisions without explicit programming. It's like teaching a child to recognize flowers by showing countless examples instead of defining rigid rules. For Hala, a machine learning algorithm can analyze historical case data, client behaviors, and court schedules to recommend the best time to prioritize specific tasks. This is not just about making educated guesses; it's about leveraging vast amounts of data to find hidden patterns and insights.

Consider a more complex scenario involving our fictional character, Rahul, a supply chain manager at a large retail company. Rahul's challenge was to predict demand for thousands of products across multiple stores, accounting for seasonal changes, regional preferences, and even social media trends. Traditional methods, which relied on explicit programming, were cumbersome and often inaccurate. They required Rahul to define every possible rule and exception, which was impractical and inefficient. With machine learning, Rahul could use algorithms to analyze past sales data, promotional impacts, and external factors like weather or local events. The system learned from this data, continuously improving its predictions over time. When a sudden spike in demand for winter jackets occurred due to an unexpected cold front, the algorithm quickly adapted, helping Rahul ensure timely restocking and preventing potential losses.

This shift from explicit programming to machine learning represents a profound change. It's about creating systems that evolve and adapt, much like humans do. Whether it's optimizing work schedules for Hala or managing complex supply chains for Rahul, machine learning unlocks new possibilities, making our tools smarter and more responsive to the ever-changing world.

There are several main types of machine learning algorithms, each suited to different types of tasks and data structures. This guide explores these algorithms, their applications, and the challenges they present.

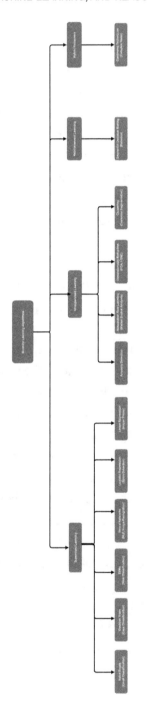

Figure 5-1. Machine learning algorithms and applications

Supervised Learning

Supervised learning algorithms train a model on labeled input data, meaning the desired output is already known. The model learns the mapping between inputs and outputs and is evaluated on its ability to predict the correct output when given new data. Common supervised learning algorithms include the following.

Linear Regression

Linear regression is used for predicting continuous values, such as estimating house prices. It learns the relationship between input features, like square footage or location, and the target variable by fitting a linear equation to the training data. Simple yet powerful, linear regression is the foundation of many predictive modeling techniques. Let's imagine Hala as our use case here; as an attorney starting at a new firm, with case patterns and client behaviors becoming increasingly erratic, Hala decided to use linear regression to predict the duration and complexity of different case types based on historical data. By understanding factors like case type, client background, and previous case outcomes, she could better allocate her time and resources, making her workflow more efficient.

Logistic Regression

Logistic regression is suited for binary classification problems like detecting spam emails or medical diagnoses. It calculates the probability of an input belonging to a particular class, then classifies it by thresholding that probability. While called logistic "regression," it actually performs classification. Hala could find logistic regression invaluable for her new role. She could use it to classify whether incoming cases were likely to settle out of court or go to trial based on features like the nature of the dispute, involved parties, and historical settlement data. This could help her prioritize her workload, focusing more on cases likely to require extensive preparation for court. She might employ logistic regression to assess the probability of winning a case based on historical data and case specifics. By calculating the probability of a particular legal outcome, logistic regression helps her make informed decisions.

Decision Trees

Decision trees break down data by making a series of binary splits based on input features. Each split leads either to a leaf node classification or another decision node. Decision trees capture nonlinear relationships and are easily interpretable, visualizing the logic behind predictions. Ensemble methods like random forests improve accuracy by combining multiple decision trees. In her quest to manage her caseload effectively, Hala could employ decision trees to identify the key factors influencing case outcomes. By visualizing these factors, she could quickly grasp which elements were most critical in determining a case's trajectory, such as the judge's history, the legal team involved, or the specifics of the legal issue at hand. Consider Hala encountering complex legal cases involving multiple variables affecting client outcomes. Decision trees allow her to visually map out the decision making process, highlighting critical factors influencing legal strategies. Ensemble methods like random forests enhance decision tree accuracy by combining multiple trees, refining predictions in intricate legal scenarios.

Support Vector Machines

Support vector machines (SVMs) find a hyperplane that best separates input data points into two classes. The hyperplane maximizes the margin between the classes to minimize generalization error. Effective in high-dimensional spaces, SVMs handle both linear and nonlinear classification using kernel tricks. They work well for complex but small or medium sized datasets. Hala might use SVMs to tackle complex classification tasks within her new specialty. For example, she could use SVMs to predict whether a case would be classified as a high-risk or low-risk case, helping her and her team allocate resources more effectively. The ability of SVMs to handle high-dimensional data was particularly useful as she dealt with cases involving numerous legal precedents and intricate details. Facing nuanced legal disputes requiring clear delineation between legal arguments, SVMs could assist in identifying pivotal legal arguments or precedents based on complex case data. Their ability to handle both linear and nonlinear classifications makes them versatile tools in legal analytics.

Neural Networks

Artificial neural networks are computing systems containing interconnected nodes that resemble neurons in the brain. They learn hierarchical feature representations from raw input data through multiple stacked layers. Deep neural networks with many hidden layers have fueled breakthroughs in computer vision, natural language processing, and other complex tasks involving high-dimensional unstructured data. In her innovative approach, Hala could leverage neural networks to analyze complex legal documents and extract relevant information. By training a neural network on a dataset of past cases, she could quickly identify crucial elements within new documents, such as pertinent laws, relevant case precedents, and potential legal strategies. This significantly speeds up her preparation process, allowing her to focus on high-level strategic decisions. Hala could leverage neural networks for tasks like document classification and contract analysis. For example, a neural network could be trained to automatically categorize legal documents based on their content, saving time and reducing the manual effort required for document review.

Naive Bayes

Naive Bayes classifiers utilize Bayes' theorem to assume input features are independent when calculating probabilities to classify data points. Their simplicity enables fast model training and inference, scaling well to large datasets. Naive Bayes works surprisingly well despite its simplifying assumptions and serves as a strong baseline classifier. To further streamline her workflow, Hala could apply Naive Bayes to quickly classify incoming emails and documents into predefined categories, such as urgent, review, or archive. This automated classification helps her prioritize her tasks and ensured that important communications were addressed promptly, maintaining the efficiency needed in her demanding legal role. She might also use a Naive Bayes classifier to detect fraudulent transactions within the firm's financial system. By analyzing patterns in transaction data, the classifier can flag suspicious activities for further investigation, enhancing the firm's financial security.

Unsupervised Learning

Unsupervised learning is a type of machine learning that uncovers hidden patterns and intrinsic structures within unlabeled input data, meaning there are no predefined categories or labels. Unlike supervised learning, which relies on labeled data with known outputs, unsupervised learning algorithms must infer the natural organization of the data independently. This approach is essential for discovering unknown patterns, grouping similar data points, and identifying anomalies. Common techniques include clustering, which groups similar items together, and dimensionality reduction, which simplifies data complexity. Key unsupervised learning techniques include clustering, dimensionality reduction, association rule learning, and anomaly detection.

Clustering

Clustering algorithms group data points together based on similarity. K-means clustering partitions data into k clusters, where each data point belongs to the cluster with the nearest mean. Hierarchical clustering builds a hierarchy of clusters using agglomerative or divisive strategies.

Clustering provides insight into data distributions and outlier detection. Hala faces the challenge of organizing vast amounts of legal data from her new cases. She could benefit from clustering algorithms to group similar cases and documents, which allowed her to identify common patterns and trends. This grouping can help her understand the typical characteristics of different types of cases and prepare more effectively by focusing on the commonalities within each cluster.

Dimensionality Reduction

Dimensionality reduction transforms high-dimensional feature spaces into lower dimensions while preserving structure. Key techniques include principal component analysis (PCA), which rotates the feature axes to capture maximal variance, and t-distributed stochastic neighbor embedding (t-SNE), which models each data point in 2D/3D space. Reducing dimensions aids visualization while removing noise. To handle the complexity of her data, Hala can utilize dimensionality reduction techniques like PCA. This allows her to visualize her data more effectively, identifying key trends and outliers without being overwhelmed by the sheer volume of information. By focusing on the most important features, she could make more informed decisions and improve her case strategies. Hala's new firm might have a vast amount of case data with numerous features. Dimensionality reduction techniques like PCA can help visualize this data in a lower-dimensional space, making it easier to identify patterns and trends that inform strategic decisions.

Association Rule Learning

Algorithms like Apriori uncover relationships between variables in large databases, quantified by support and confidence metrics. Association rules connect antecedent and consequent variable values, like "90% of customers who purchased item A also purchased item B." This reveals product bundles for market basket analysis in retail. Hala can use association rule learning to identify common co-occurrences in her case data. For example, she might discover that certain legal issues often appeared together or that specific types of clients were more likely to have particular legal needs. This knowledge can help her anticipate client needs better and tailor her services accordingly. By uncovering these associations, the firm can refine its approach to increase the likelihood of favorable outcomes.

Anomaly Detection

Anomaly or outlier detection finds rare data instances that differ significantly from the majority distribution. Identifying anomalies enables detecting fraud, network intrusions, faults in systems, and more. Techniques range from simple statistical metrics to isolation forests and neural networks. In her practice, Hala encountered the need to identify unusual cases that might require special attention. She might employ anomaly detection techniques to flag cases that deviated significantly from the norm, such as those with unusually high stakes or complex legal issues. This proactive approach ensures that no critical details were overlooked. She might use anomaly detection to monitor the firm's cybersecurity. By identifying unusual patterns of network activity, the system can flag potential security breaches, allowing for timely intervention and protection of sensitive data.

Reinforcement Learning

Reinforcement learning trains AI agents to make optimal decisions sequentially in dynamic environments. The agent learns to maximize cumulative future reward through trial and error interactions with its environment. Key reinforcement learning techniques include Q-learning, deep Q-networks, policy gradients, and multi-armed bandits.

Q-learning is a model free reinforcement learning technique that estimates the long-term value or quality of actions based on expected future rewards. It balances the exploration of uncharted territory with exploitation of known rewards. For example, Hala could use Q-learning to optimize her firm's resource allocation. By evaluating the long-term value of different resource allocation strategies, the system can recommend actions that maximize the firm's overall efficiency and effectiveness.

The Future of Machine Learning

Machine learning powers a vast range of AI applications today. As algorithms grow more advanced, learning becomes more efficient, expanding use cases across industries. Hala's journey demonstrates how machine learning can transform a traditionally human centric field like law, providing tools that enhance decision making, streamline workflows, and ultimately improve outcomes for clients. Continued progress calls for addressing emerging issues like bias in data and models as well as the environmental impacts of power hungry deep learning. However, machine learning provides the vehicle for AI to drive monumental technological change now and into the future.

Data serves as the lifeblood of machine learning systems. Understanding the different data types and formats, mastering the preprocessing techniques that transform raw data into informative features, and prioritizing data quality are fundamental to extracting actionable insights. Just as quality ingredients are essential for creating a gourmet meal, high-quality data forms the cornerstone for advanced algorithms to uncover valuable patterns and trends in complex datasets.

Reasoners in AI

The digital world buzzes with data, algorithms crunch numbers at breakneck speeds, and machine learning models churn out predictions. But in artificial intelligence, there's a crucial component that elevates these systems from mere number crunchers to something approaching true intelligence: Reasoners.

Imagine our friend Hala, the attorney we met earlier, facing a complex legal case. She's drowning in precedents, statutes, and case files. While her machine learning models can categorize documents and predict case outcomes, they lack the ability to connect the dots, to see the bigger picture. This is where reasoners come into play. Reasoners in AI are like the logical mind of a seasoned lawyer or the analytical brain of a chess grandmaster. They take the raw information, the patterns identified by machine learning, and apply logical rules to derive conclusions. It's not just about recognizing patterns, it's about understanding implications, making deductions, and arriving at well-reasoned decisions.

For Hala, a well-designed AI reasoner could be a game-changer. It could analyze the nuances of her case, consider relevant precedents, and construct logical arguments. More than that, it could explain its reasoning, providing Hala with insights she might have overlooked. But what exactly are these reasoners, and how do they work? Let's dive deeper into the AI reasoning systems.

Rule-Based Reasoners

At their core, rule-based reasoners are like the strict, by-the-book judges of the AI world. They operate on a set of predefined if–then rules, much like the legal codes Hala relies on in her practice. These systems excel in domains where expert knowledge can be clearly articulated and formalized.

Imagine Hala working on a tax law case. A rule-based reasoner might have rules like:

- If (income > $500,000) AND (filing_status = single) THEN (tax_bracket = 37%)

- If (deductions > standard_deduction) THEN (filing_method = itemized)

These rules, when combined and applied to specific case details, can quickly navigate through complex tax regulations, helping Hala determine the correct tax treatment for her client. The beauty of rule-based systems lies in their transparency. Every decision can be traced back to specific rules, making it easy for Hala to explain her recommendations to clients or justify her arguments in court. It's like having a tireless legal assistant who never forgets a statute or misinterprets a regulation.

However, as any experienced lawyer knows, real-world cases rarely fit neatly into predefined rules. What happens when a case falls into a gray area? Or when new legislation introduces ambiguity? This is where rule-based systems start to show their limitations. They can be brittle, struggling with scenarios not explicitly covered by their rule set. As Hala's practice grows more complex, she might find herself constantly updating and refining the rules, a task that quickly becomes overwhelming.

Probabilistic Reasoners

Enter probabilistic reasoners, the more flexible cousins of rule-based systems. These reasoners acknowledge what every good lawyer knows: The world is full of uncertainty. Instead of dealing in absolutes, they work with probabilities, weighing evidence and making informed judgments.

For Hala, a probabilistic reasoner could be invaluable in assessing the strength of her cases. It might consider factors like:

- The judge's historical rulings on similar cases

- The reliability of witnesses

- The persuasiveness of available evidence

Using techniques like Bayesian networks, the system could calculate the probability of various outcomes, helping Hala make strategic decisions about how to proceed.

Let's say Hala is considering whether to push for a settlement or go to trial. A probabilistic reasoner might analyze:

- The strength of the evidence (70% favorable)

- The opponent's history of settling (60% likely to settle)

- The potential damages if the case goes to trial (mean estimate $500,000, with high variance)

Based on these factors and more, the system might recommend pursuing settlement negotiations, estimating a 65% chance of achieving a favorable outcome without the risks and costs of a full trial.

The power of probabilistic reasoners lies in their ability to handle nuance and uncertainty, much like an experienced lawyer weighing the subtleties of a case. They can update their beliefs as new evidence emerges, mirroring the way a skilled attorney adjusts strategy as a case unfolds. However, these systems are not without challenges. The accuracy of their predictions depends heavily on the quality of the probability estimates fed into them. In the ever-changing legal landscape, obtaining reliable probabilities can be a formidable task. Moreover, while probabilistic reasoners can provide valuable insights, explaining their reasoning to clients or in court can be more challenging than with simpler rule-based systems.

Logical Reasoners

While rule-based and probabilistic reasoners handle many practical tasks, some legal arguments require a higher level of abstraction. This is where logical reasoners shine. Based on formal logic systems, these reasoners can handle complex, abstract reasoning tasks that mirror the kind of high-level thinking required in constitutional law or in arguing novel legal theories.

For Hala, a logical reasoner might help in constructing arguments based on legal principles and precedents. Using formalisms like first order logic, the system could represent complex legal concepts and their relationships. For instance, it might formalize the concept of "due process" and its implications across different types of cases.

Logical reasoners excel at tasks like:

- Analyzing the logical consistency of legal arguments

- Identifying potential counterarguments

- Exploring the implications of new precedents on existing law

Imagine Hala working on a groundbreaking civil rights case. A logical reasoner could help her explore the ramifications of various legal arguments, ensuring her approach is logically sound and considering potential challenges from opposing counsel.

The strength of logical reasoners lies in their ability to handle abstract concepts and complex relationships, much like the analytical mind of a top tier lawyer. They can uncover subtle logical flaws in arguments or identify novel legal strategies by exploring the logical consequences of different legal principles.

Like their human counterparts, logical reasoners can sometimes get lost in abstraction. They may struggle with the messiness of real-world data and the nuances of human behavior that often play important roles in legal outcomes. Moreover, translating their abstract logical constructs into persuasive arguments that resonate with judges and juries remains a challenge.

Hybrid Reasoners

As Hala's practice grows more complex, she realizes that no single approach to reasoning can handle all the challenges she faces. This realization mirrors a trend in AI: the development of hybrid reasoners that combine multiple reasoning techniques. Hybrid reasoners aim to leverage the strengths of different approaches while mitigating their individual weaknesses. For Hala, a hybrid system might combine:

- Rule-based reasoning for straightforward regulatory compliance checks

- Probabilistic reasoning for case outcome prediction and risk assessment

- Logical reasoning for constructing and analyzing complex legal arguments

- Machine learning for pattern recognition in large volumes of case law

Such a system could provide Hala with a powerful suite of tools, adapting its approach based on the specific needs of each case. For a routine contract review, it might rely heavily on rule based components. For a complex litigation strategy, it could employ a combination of probabilistic assessment and logical argument construction. The power of hybrid systems lies in their flexibility and comprehensiveness. They can handle a wide range of legal tasks, from the mundane to the highly complex, mirroring the versatility required of successful lawyers.

However, with great power comes great complexity. Hybrid systems can be challenging to develop, maintain, and explain. Ensuring that different reasoning components work together seamlessly and don't produce conflicting advice is an ongoing challenge in AI research. As Hala looks to the future of her practice, she sees both exciting possibilities and daunting challenges in the world of AI reasoners. The potential for these systems to augment and enhance legal practice is immense. They could help lawyers like Hala handle larger caseloads, uncover insights in vast legal databases, and construct more robust arguments. Yet, significant hurdles remain. Developing AI systems that can truly understand the nuances of legal language, the complexities of human behavior, and the ever-evolving nature of law is a formidable challenge. Moreover, questions of accountability and explainability loom large. How do we ensure that AI-assisted legal decisions are fair, transparent, and accountable?

As researchers work to address these challenges, the future of AI reasoners in law looks bright. We might see systems that can:

- Automatically generate legal briefs, adapting their arguments based on the specific judge assigned to the case

- Predict shifts in legal interpretations by analyzing trends in court decisions across jurisdictions

- Assist in real time during trials, providing instant access to relevant precedents and suggesting lines of questioning

For Hala and her colleagues, these advancements promise to revolutionize legal practice. But they also raise profound questions about the nature of legal expertise and the role of human judgment in the law.

As we conclude our exploration of reasoners in AI, it's clear that these systems represent more than just technological tools. They embody our attempts to codify human reasoning, to capture the essence of logic and decision-making in digital form. From the rule-based systems that mirror our most straightforward thought processes to the complex hybrid reasoners that juggle probabilities and abstract logic, each type of reasoner reflects a different facet of human cognition.

Yet, as powerful as these systems become, they remain tools – extraordinarily sophisticated tools, but tools nonetheless. The true power lies in the synergy between human expertise and AI capabilities. Hala's success in navigating complex legal waters will depend not just on the capabilities of her AI assistants, but on her ability to wield these tools effectively, to know when to rely on their insights and when to trust her own instincts.

As we look to the future, the development of AI reasoners will likely continue to mirror our understanding of human cognition. Advances in fields like cognitive science and neuroscience may inform new approaches to AI reasoning, creating systems that more closely mimic the flexibility and intuition of human thought.

At the same time, AI reasoners may push the boundaries of what we consider possible in logical analysis and decision-making. They may uncover patterns and connections that human minds overlook, leading to new insights in law, science, and beyond. For professionals like Hala, the key will be to embrace these tools while maintaining a critical perspective. Understanding the strengths and limitations of different reasoning approaches, knowing how to interpret and explain their outputs, and recognizing when human judgment should override AI recommendations will be crucial skills in the AI-augmented workplaces of the future.

As we close this chapter, we're reminded that the story of AI reasoners is, at its heart, a story about human reasoning. It's about our attempts to understand our own thought processes, to codify our logic, and to extend our cognitive abilities. Whether in law, medicine, finance, or any field that requires complex decision making, AI reasoners stand poised to become invaluable partners in human endeavors.

For Hala, as she faces her next challenging case, her AI reasoning systems won't replace her legal expertise–they'll enhance it. They'll help her see connections she might have missed, consider angles she might have overlooked, and construct arguments more robust than ever before. But the final decisions, the ethical judgments, and the human touch that clients seek–those will remain firmly in her capable hands. The future of AI reasoners is not about machines replacing human thought, but about creating a powerful synergy between human and artificial intelligence. It's a future where the boundaries of what's possible in reasoning and decision-making are constantly expanding, driven by the combined power of human creativity and machine precision. As we look ahead, the potential applications of AI reasoners seem boundless. From helping solve global challenges like climate change and resource allocation to assisting in personal decision-making in our daily lives, these systems will likely become an integral part of how we approach complex problems.

Conclusion

With that being said, as with all powerful technologies, the rise of AI reasoners will require careful consideration of ethical implications. How do we ensure that these systems are used responsibly? How do we maintain human autonomy in decision-making while leveraging the power of AI? These are questions that society as a whole will need to grapple with as AI reasoners become more prevalent and powerful. We will talk more about legal and ethical issues within your AI solutions in a separate chapter. For professionals like Hala, and indeed for all of us, the coming years will be a time of adaptation and learning. We'll need to develop new skills to work effectively with AI reasoners, to understand their outputs, and to integrate their insights into our decision-making processes. It's a challenge, certainly, but also an exciting opportunity to expand the boundaries of human knowledge and capability.

As we conclude, let's remember that the story of AI reasoners is still being written. Each new development, each new application, adds a new chapter to this unfolding narrative. It's a story that intertwines the realms of computer science, cognitive psychology, philosophy, and countless other disciplines. And it's a story in which we all play a part, as we shape the future of artificial intelligence and its role in our world.

For Hala, and for all of us, the journey with AI reasoners is just beginning. It promises to be a journey of discovery, challenge, and unprecedented possibility. As we embark on this path, we carry with us the age old human qualities of curiosity, creativity, and critical thinking, qualities that will remain essential, no matter how advanced our AI companions become. In the next chapter, we will talk about large language models (LLMs), and to illustrate the transformative power of LLMs, we will follow the journey of Sarah, the innovative CEO of TechStyle, a mid-sized imaginary online fashion retailer.

CHAPTER 6

Large Language Models (LLMs)

Introduction

In this chapter, we'll explore the evolution, capabilities, and applications of LLMs, with a particular focus on how they are revolutionizing ecommerce and retail automation. Through the story of Sarah, the innovative CEO of TechStyle, a mid-sized imaginary online fashion retailer, we'll illustrate the practical implications of LLMs in business. We'll delve into the technical aspects of LLMs, their ability to understand and generate human-like text, and their emergent reasoning capabilities. Additionally, we'll examine the future potential of LLMs, including their integration with other emerging technologies, and the ethical considerations surrounding their development and use.

What Are Large Language Models (LLMs)?

The story of LLMs is one of rapid transformation. It begins in the early days of natural language processing, where simple rule-based systems struggled to make sense of the complexities of human communication. These early attempts, while groundbreaking for their time, were limited in their ability

© Anton Cagle, Ahmed Mohamed Ceifelnasr Ahmed 2024
A. Cagle and A. M. C. Ahmed, *Architecting Enterprise AI Applications*,
https://doi.org/10.1007/979-8-8688-0902-6_6

to truly understand context and nuance. They were like children learning to read, laboriously sounding out each word without grasping the meaning of the sentence as a whole.

Sarah remembers these days well. TechStyle's first attempt at automation involved a simple chatbot that could only respond to a handful of predefined queries. Customers found it frustrating, and it did little to ease the burden on TechStyle's human customer service team. As we progressed through the latter part of the 20th century, statistical models began to show promise. These approaches, relying on the frequency and co-occurrence of words, allowed for more sophisticated analysis of text. Yet, they still fell short of true understanding, often producing results that were statistically sound but semantically nonsensical. The real revolution came with the advent of neural networks and deep learning. Suddenly, machines could process language in ways that mimicked the human brain, learning patterns and relationships that went beyond simple word associations. Models like Word2Vec and GloVe introduced the concept of word embeddings, representing words as points in a multidimensional space where proximity indicated semantic similarity. It was as if we had given machines a map of language, allowing them to navigate the terrain of meaning with increasing dimentionality.

But the true game-changer arrived in 2017 with the introduction of the Transformer architecture. This ingenious design, with its attention mechanism allowing for parallel processing of input sequences, unlocked new ways of possibility in language modeling. It was like giving our AI linguists a pair of multifaceted glasses, enabling them to focus on different parts of a sentence simultaneously, weighing the importance of each word in relation to all the others. The Transformer architecture paved the way for models like BERT and the GPT series, each pushing the boundaries of what was possible in language understanding and generation. These models, trained on vast corpora of text encompassing everything from classic literature to scientific papers to social media posts, began to exhibit capabilities that seemed almost magical.

LLM Applications and Capabilities

Large language models (LLMs) have revolutionized various applications with their ability to understand and generate human-like text. At TechStyle, they implemented an LLM-powered chatbot capable of handling complex customer inquiries with remarkable accuracy, maintaining context, and even detecting emotional cues. Beyond customer service, LLMs have been utilized for personalized content generation, such as creating tailored product descriptions that significantly boosted conversion rates. Furthermore, LLMs demonstrated emergent reasoning abilities, enabling TechStyle to develop a virtual personal stylist that curates fashion recommendations based on nuanced customer preferences. These capabilities have also enhanced TechStyle's inventory management by predicting fashion trends through comprehensive data analysis, showcasing the transformative potential of LLMs across industries.

TechStyle's LLM Experience

For Sarah and TechStyle, this breakthrough was a revelation. They implemented an LLM-powered chatbot on their website and mobile app. This wasn't your average chatbot; it could understand and respond to complex customer queries with remarkable accuracy and human-like reasoning. Whether a customer was asking about the fit of a particular dress, the sustainability of TechStyle's manufacturing processes, or needed help with a return, the AI could handle it all. What impressed Sarah most was the chatbot's ability to maintain context over long conversations, even picking up on subtle emotional cues in customer messages. For instance, when a customer expressed frustration about a delayed order, the AI not only provided tracking information but also offered a sincere apology and a discount on their next purchase–all in a tone that matched the customer's mood. Imagine, if you will, a machine that can not only understand the literal meaning of words but can grasp context, detect sarcasm, and even engage in creative writing. This is the world of modern LLMs. The architecture of these models is an innovation of engineering. At their basics, LLMs are massive neural networks, often containing hundreds of billions of parameters. These parameters, adjusted through the process of training, encode the patterns and relationships found in language. The embedding layer serves as the entry point, converting words or subwords into dense vector representations that capture semantic relationships. It's as if each word is given a unique fingerprint that encodes its meaning and usage. As we dig deeper into the architecture, we encounter the multihead attention mechanism, the true genius of the Transformer design. This component allows the model to weigh the importance of different words in the input when processing each word, enabling it to capture complex relationships and dependencies within the text. It's similiar to a skilled reader who can instantly recognize which parts of a sentence are important to its meaning and which are merely supplementary. The feed-forward neural networks that process the output of the attention layers allow for nonlinear transformations of the data, adding depth and complexity to the model's understanding. Layer normalization and residual connections act as stabilizing forces, allowing these deep networks to be trained effectively.

When an LLM generates text, it does so one token at a time, each choice informed by the vast network of probabilities and relationships it has learned. A token can be as small as a single character or as large as an entire word or phrase, depending on the language and the model's tokenization scheme. It's a process that can produce remarkably coherent and contextually appropriate text, often indistinguishable from human-written prose.

Sarah's team at TechStyle leveraged this capability to revolutionize their product descriptions. Instead of generic, one-size-fits-all text, each item on TechStyle's website now had a unique, engaging description. The LLM could generate these descriptions based on product specifications, taking into account current fashion trends, and even tailoring the language to appeal to different customer segments. For example, a simple black dress might be described as "A versatile classic for the office warrior" to a professional woman in her 30s, while the same dress could be presented as "The perfect canvas for your wildest accessorizing dreams" to a fashion-forward college student. This level of personalization led to a significant increase in conversion rates across all product categories.

Reasoning with LLMs

But perhaps the most fascinating aspect of LLMs is their emergent ability to perform reasoning tasks. Despite not being explicitly designed for logical inference, these models have demonstrated an uncanny knack for problem-solving and analytical thinking. This capability arises from the patterns and relationships learned during training, allowing LLMs to recognize the structure of logical arguments, identify causal relationships, and apply problem-solving approaches they've encountered in their vast training data. When faced with a reasoning task, an LLM processes the entire context of the given prompt or question, grasping sense and specifics that might elude simpler systems. The model can break down problems into steps, maintain coherence across multiple stages of reasoning, and even draw analogies between different concepts or situations. This ability to reason extends to natural language understanding tasks as well. LLMs excel at named entity recognition, effortlessly identifying and classifying entities like person names, organizations, and locations within text. They can perform sentiment analysis with remarkable accuracy, discerning the emotional tone of a piece of text and even detecting subtle cues like sarcasm or irony. They can navigate the complexities of semantic role labeling, understanding who did what to whom, when, where, and why in a given sentence.

For TechStyle, this reasoning capability opened up exciting new possibilities. Sarah's team developed a virtual personal stylist, powered by the same LLM technology. Customers could chat with this AI stylist about their preferences, body type, upcoming events, and even send photos of their existing wardrobe. The AI would then curate personalized outfit recommendations, mixing and matching items from TechStyle's inventory. What set this virtual stylist apart was its ability to engage in nuanced fashion discussions. It could explain why certain colors complemented a customer's skin tone, suggest accessories that would elevate an outfit, and even offer styling tips for different occasions. The AI stylist became so popular that TechStyle saw a 20% increase in average order value among users who engaged with it. As we continue to explore and refine these models, we're uncovering new capabilities and pushing the boundaries of what's possible in artificial intelligence. LLMs are finding applications in diverse fields, from healthcare to finance to education, revolutionizing how we interact with and process information. Yet, for all their impressive abilities, it's important to remember that LLMs are not conscious entities. They don't truly understand language in the way humans do but rather excel at recognizing and reproducing patterns. Their outputs, while often remarkably human like, are fundamentally based on statistical patterns in their training data rather than true comprehension or reasoning.

Sarah was acutely aware of this limitation. While the LLM powered systems at TechStyle were incredibly effective, she ensured that her team understood that these were tools to augment human creativity and decision making, not replace them entirely. This perspective was needed in maintaining the human touch that had always been a cornerstone of TechStyle's brand. As we look to the future, the potential of LLMs seems boundless. The challenges that lie ahead, from improving efficiency and reducing model size to enhancing capabilities in areas like multimodal learning and long-term memory, promise to keep researchers and developers busy for years to come.

In the following sections, we'll delve deeper into specific aspects of LLMs, exploring their applications, limitations, and the ethical considerations that arise as these powerful tools become increasingly integrated into our daily lives. The story of LLMs is far from over; indeed, it feels as though we've only just begun to scratch the surface of what's possible in the world of artificial intelligence and language.

Take, for instance, their prowess in logical reasoning. These models, trained on vast troves of human knowledge, have developed a huge ability to navigate complex logical problems with a finesse that often rivals human experts. More impressively, it can apply this understanding to novel situations, demonstrating a form of deductive reasoning that extends far beyond mere pattern matching. But the true marvel lies in how LLMs handle more complex, real-world reasoning tasks. Consider a scenario where TechStyle's LLM is tasked with analyzing customer feedback and sales data to identify trends and make inventory decisions. It doesn't just tally up numbers, it can interpret the sentiment behind customer reviews, correlate it with sales figures, and even consider external factors like social media trends or economic indicators to make nuanced predictions about future demand. The way LLMs approach these tasks is fundamentally different from traditional AI systems. Rather than following a set of predefined rules, they draw upon a vast network of interconnected knowledge, weighing probabilities and making inferences in a manner that, at times, feels almost intuitive.

At TechStyle, this capability transformed their inventory management and trend forecasting. By analyzing vast amounts of data–from social media trends and fashion blog posts to historical sales data and customer feedback–the AI could predict upcoming fashion trends with accuracy. This allowed TechStyle to optimize its inventory, ensuring they always had the right products in stock at the right time.

The Future of LLMs: Potential, Implications, and Responsible Development

The future of large language models (LLMs) holds huge potential, promising to revolutionize industries through their advanced capabilities. However, as these models evolve, they also bring significant implications for society, particularly in areas like privacy, data protection, and the potential for misuse, such as generating misinformation. Responsible development is crucial to mitigate these risks, requiring robust governance frameworks and ethical oversight. The integration of LLMs with emerging technologies like quantum computing and multimodal AI could further expand their impact, but this necessitates careful consideration of the philosophical and societal challenges posed by increasingly sophisticated AI systems.

Limitations and Ethical Considerations

For all their impressive capabilities, LLMs are not infallible. They can fall prey to biases present in their training data, sometimes producing outputs that reflect societal prejudices or historical inaccuracies. They can also generate plausible sounding but factually incorrect information, a phenomenon often referred to as "hallucination" in AI circles. These limitations serve as a stark reminder that while LLMs are powerful tools, they require careful oversight and validation, especially in high stakes applications.

Sarah and her team at TechStyle encountered these challenges firsthand. There were occasional hiccups–like when the AI stylist recommended winter coats to customers in tropical climates, or when it generated product descriptions with unintentionally humorous metaphors. But Sarah saw these as learning opportunities, continuously fine-tuning the AI to improve its performance. How do we ensure these models are used responsibly? What are the implications for privacy and data protection? How do we address the potential for misuse, such as the generation of convincing misinformation or deepfakes? These are not just technical challenges but societal ones that require thoughtful consideration and robust governance frameworks. For TechStyle, this meant implementing strict data privacy measures and being transparent with customers about how their information was being used. Sarah also established an ethics board to oversee the company's AI initiatives, ensuring that the technology was being deployed in ways that aligned with TechStyle's values and respected customer rights. The potential applications of LLMs extend far beyond text generation and analysis.

We're seeing them employed in creative endeavors, assisting writers and artists in generating novel ideas and content. In scientific research, they're helping to sift through vast amounts of literature, identifying patterns and connections that human researchers might overlook. In education, they're being used to create personalized learning experiences, adapting to each student's unique needs and learning style.

At TechStyle, the creative potential of LLMs was particularly exciting. The AI began to assist in designing new clothing lines, generating ideas based on current trends, customer preferences, and even sustainability considerations. While human designers still played a great role in the process, the AI's ability to quickly generate and iterate on ideas accelerated the design process significantly. One particularly exciting frontier is the integration of LLMs with other AI technologies. Imagine combining the language understanding capabilities of an LLM with a computer vision system, creating an AI that can not only see the world but describe and reason about what it sees in natural language. Or consider the potential of pairing LLMs with robotic systems, enabling machines that can understand complex verbal instructions and carry out physical tasks with precision. Sarah was already exploring these possibilities at TechStyle. She envisioned AI-powered virtual fashion shows where customers could see clothes on models that looked like them, with the LLM providing detailed, personalized commentary on each outfit. Or LLM-generated fashion blogs that could produce an endless stream of style inspiration tailored to each reader's tastes. Researchers are exploring ways to make these models more efficient, reducing their computational requirements while maintaining or even improving their performance. There's also a push toward more interpretable models, seeking to shed light on the often decision making processes of these complex systems. Another area of active research is in developing LLMs that can learn and adapt in real time, updating their knowledge base as they interact with the world. This could lead to AI assistants that grow and evolve alongside their users, becoming increasingly personalized and helpful over time.

For TechStyle, this could mean an AI stylist that not only recommends outfits but learns from each customer's feedback and purchasing decisions, continually refining its understanding of their style preferences. It could also lead to more dynamic pricing strategies, with the AI adjusting prices in real time based on demand, inventory levels, and even individual customer behavior. The impact of LLMs on society is likely to be profound and far-reaching. As these systems become more sophisticated and widely deployed, they have the potential to revolutionize how we interact with technology, access information, and even how we think about intelligence itself. They could democratize access to knowledge, breaking down language barriers, and making vast amounts of information accessible to anyone with an Internet connection. However, this technological revolution also raises important questions about the future of work and education. As LLMs become capable of performing tasks that were once the exclusive domain of human professionals, how will job markets and educational systems adapt? Will we see a shift toward more creative and interpersonal skills that are harder for AI to replicate? Or will we find new ways to collaborate with these AI systems, enhancing rather than replacing human capabilities? Sarah grappled with these questions at TechStyle. While the AI systems had dramatically improved efficiency and customer experience, she was committed to ensuring that her human employees weren't left behind. She invested in training programs to help her team develop skills that complemented the AI, focusing on areas like creative strategy, ethical decision-making, and advanced data interpretation. Large language models represent one of the most exciting and potentially transformative technologies of our time. They embody the dream of AI that can understand and communicate in human language, opening up new possibilities for human–machine interaction and collaboration. As we continue to explore and develop these powerful tools, we must remain mindful of both their immense potential and the responsibility that comes with wielding such powerful technology.

The story of LLMs is still being written, with each new breakthrough opening up new questions and possibilities. As Sarah reflected on TechStyle's AI-driven transformation, she realized that LLMs had not just improved their operations, they had fundamentally changed how they connected with customers. The technology allowed TechStyle to provide a level of personalization and service that was previously impossible at scale, blurring the lines between online and in-store shopping experiences

The integration of LLMs with other cutting-edge technologies promises to usher in a new era of artificial intelligence that could fundamentally transform our world. Consider, for instance, the potential of combining LLMs with quantum computing. While still in its infancy, quantum computing offers the promise of exponentially increased processing power. If harnessed for language models, we could see AI systems capable of processing and analyzing language at speeds and scales that are currently unimaginable. This could lead to real-time language translation for any language pair, instantaneous analysis of global communication patterns, or even the ability to process and understand the entirety of human written knowledge in mere moments. For TechStyle, this could mean the ability to analyze global fashion trends in real time, predicting and even setting trends before they emerge. It could also enable the creation of truly personalized shopping experiences, where every aspect of the customer's interaction with the brand is tailored to their individual preferences and needs.

Integrating LLMs with Emerging Technologies

Another exciting frontier is the development of multimodal AI systems that seamlessly integrate language understanding with other forms of perception. Imagine an AI that can not only understand and generate language but can also see, hear, and even interact with the physical world. Such systems could revolutionize fields like robotics, where machines could understand complex verbal instructions and carry them out with human-like dexterity. In the context of ecommerce and retail, this could lead to virtual shopping assistants that can see and understand a customer's existing wardrobe, body type, and personal style, offering highly accurate and personalized fashion advice. It could also enable more sophisticated augmented reality applications, allowing customers to virtually try on clothes and see how they look in different settings. The potential applications of advanced LLMs in scientific research are particularly thrilling.

We're already seeing AI systems that can read and synthesize scientific literature, but future models might be able to generate and test hypotheses, design experiments, and even make scientific discoveries. This could accelerate the pace of scientific progress dramatically, potentially leading to breakthroughs in fields like medicine, climate science, and physics. For TechStyle, this could translate into AI-driven innovation in textile science and sustainable fashion. Imagine an LLM that could analyze research on eco-friendly materials, consumer preferences, and manufacturing techniques to suggest novel, sustainable fabric blends or innovative, zero waste clothing designs.

Philosophical and Societal Implications

However, as we push the boundaries of what's possible with LLMs, we must also grapple with increasingly complex ethical and philosophical questions. As these models become more sophisticated, the line between artificial and human intelligence may begin to blur. We may find ourselves asking profound questions about the nature of consciousness, creativity, and even what it means to be human. Sarah and her team at TechStyle were acutely aware of these challenges. They grappled with questions like: At what point does AI generated fashion design infringe on human creativity? How do we ensure that AI driven personalization doesn't create echo chambers, limiting customers' exposure to new styles and ideas?

These were not easy questions to answer, but Sarah knew that addressing them head-on was crucial for the ethical deployment of AI in their business.

There are also pressing concerns about the societal impacts of widespread LLM deployment. How do we ensure equitable access to these powerful tools? How do we prevent the exacerbation of existing social inequalities? And how do we safeguard against the potential misuse of LLMs for surveillance, manipulation, or the spread of misinformation? The development of AI governance frameworks will be very important in addressing these challenges. We'll need to strike a delicate balance between fostering innovation and ensuring responsible development and deployment of these technologies. This will likely require collaboration between technologists, ethicists, policymakers, and representatives from diverse communities to ensure that the benefits of LLMs are broadly shared and potential harms are mitigated. At TechStyle, this meant not only complying with data protection regulations but going above and beyond to ensure ethical AI use. Sarah established an AI ethics board, comprising not just company executives but also ethicists, customer advocates, and even critics of AI in fashion. This board helped shape TechStyle's AI policies, ensuring that their use of LLMs and other AI technologies aligned with their values and societal expectations.

Education systems will need to evolve to prepare future generations for a world where AI is evolving. This might involve a shift toward emphasizing uniquely human skills like creativity, emotional intelligence, and ethical reasoning, areas where humans are likely to maintain an edge over AI for the foreseeable future. At the same time, we'll need to ensure widespread AI literacy, enabling people to understand, interact with, and critically evaluate AI systems. Sarah recognized this need within her own organization. She initiated partnerships with local universities to develop courses on AI in fashion and retail, helping to cultivate a workforce that could thrive in this new AI-driven landscape. She also invested in ongoing training for her current employees, ensuring they had the skills to work alongside AI systems effectively. She also prepared internship programs to leverage the ideas of college students and help them understand the impact of AI and how it's shaping industries.

The future of LLMs, and indeed of AI as a whole, is not predetermined. It will be shaped by the choices we make today, in research priorities, in ethical guidelines, in governance structures, and in how we choose to integrate these technologies into our lives and societies. The decisions we make will echo through the ages, potentially shaping the very course of human civilization. For Sarah and TechStyle, the journey with LLMs was just beginning. As she looked out over the bustling office, where AI systems and human creativity worked in harmony to push the boundaries of fashion and technology, she felt a sense of excitement and purpose. She knew that the path ahead would be challenging, filled with technical hurdles, ethical dilemmas, and unforeseen obstacles. But she also knew that with careful stewardship, a commitment to ethics, and a spirit of innovation, they could help shape a future where AI and human ingenuity combined to create something truly extraordinary. The story of TechStyle is just one example of how LLMs are transforming industries and reshaping our world. From healthcare to education, from scientific research to creative arts, these powerful AI systems are opening up new possibilities and challenging our preconceptions about the limits of machine intelligence.

LLM Application in Ecommerce

In the fast world of ecommerce, staying ahead of customer expectations is key. Sarah, the innovative CEO of TechStyle, who once again leads the charge in transforming online retail with the power of large language models (LLMs). Sarah's vision for TechStyle includes leveraging LLMs to not only enhance customer interactions but also to drive sales and personalize the shopping experience in unprecedented ways. This figure shows an LLM application in ecommerce website.

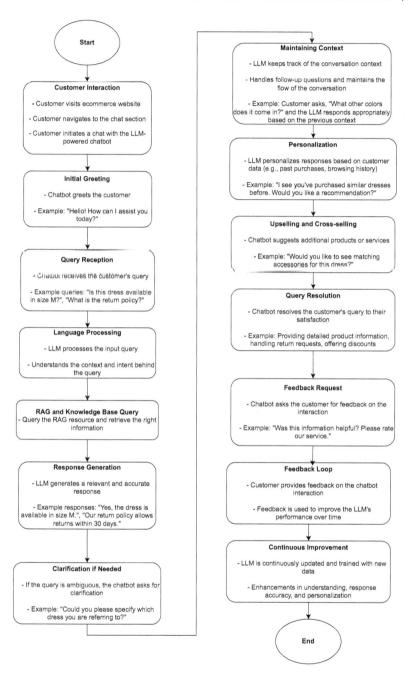

Figure 6-1. *A flowchart of an LLM application in an ecommerce website*

The Flow of Interaction

Sarah's latest project revolves around integrating an LLM-powered chatbot into TechStyle's ecommerce platform. Let's follow the journey of a customer, Shorouk, as she navigates this new and improved shopping experience.

Customer Interaction

Shorouk visits TechStyle's website on a quiet evening, ready to find a new outfit for an upcoming event. As she browses, she notices a chat icon inviting her to ask for assistance. Intrigued, she clicks it and initiates a chat with the LLM-powered chatbot.

Initial Greeting

The chatbot greets Shorouk warmly: "Hello, Shorouk! How can I assist you today?" This personalized touch immediately makes Shorouk feel valued. The greeting isn't just generic; it includes her name, showcasing the LLM's ability to create a more engaging and individualized customer experience.

Query Reception

Shorouk asks, "Is this dress available in size M?" The chatbot receives her query and processes it efficiently. The efficiency and speed of the response are critical in maintaining Shorouk's interest and trust in the system.

Language Processing

Behind the scenes, the LLM analyzes Shorouk's question, understanding both the context and the intent. It's not just about recognizing words but about grasping the meaning and urgency behind them. The ability to understand context is what sets LLMs apart from traditional chatbots, enabling them to provide more accurate and relevant responses.

Response Generation

In seconds, the chatbot responds: "Yes, the dress is available in size M. Would you like to add it to your cart?" The accuracy and speed of the response impress Shorouk. This immediate and precise interaction highlights the effectiveness of LLMs in handling common customer inquiries seamlessly.

Clarification if Needed

When Shorouk follows up with, "What other colors does it come in?" the chatbot seamlessly continues, "The dress is available in red, blue, and black. Which one would you like to see?" The chatbot's ability to handle follow-up questions without losing context demonstrates its sophisticated conversational skills.

Maintaining Context

Throughout the conversation, the chatbot maintains context, making it easy for Shorouk to ask follow-up questions without repeating herself. For instance, when Shorouk asks about matching accessories, the chatbot suggests, "We have a beautiful set of earrings that would go perfectly with the blue dress." This ability to retain context makes the interaction feel more natural and fluid, akin to conversing with a human assistant.

Personalization

Leveraging data from Shorouk's past purchases and browsing history, the chatbot personalizes its recommendations: "I see you've bought similar dresses before. Would you like a recommendation for shoes to match?" By personalizing responses based on Shorouk's preferences and shopping history, the chatbot enhances her shopping experience, making it more tailored and enjoyable.

Upselling and Cross-Selling

To maximize the shopping experience, the chatbot suggests additional products: "Would you like to see matching accessories for this dress?" This not only enhances Shorouk's experience but also boosts TechStyle's sales. Effective upselling and cross-selling can significantly increase the average order value, benefiting both the customer and the business.

Query Resolution

The chatbot ensures Shorouk's query is fully resolved, providing detailed information and even handling a return request when needed. "If you're not satisfied with the dress, our return policy allows returns within 30 days." Ensuring comprehensive query resolution helps build customer trust and loyalty, essential for long-term success in ecommerce.

Feedback Request

Before concluding the chat, the chatbot asks for feedback: "Was this information helpful? Please rate our service." Shorouk feels her opinion is valued, closing her shopping session on a positive note. Gathering feedback is crucial for continuous improvement and for understanding customer satisfaction.

Feedback Loop

Shorouk's feedback is then analyzed to continuously improve the chatbot's performance. The LLM learns from each interaction, enhancing its responses for future customers. This feedback loop is integral to maintaining high standards of customer service and adapting to evolving customer needs.

Continuous Improvement

TechStyle's LLM is regularly updated with new data, ensuring it remains accurate and relevant. This continuous improvement cycle is vital in maintaining high customer satisfaction. Regular updates and training help the LLM adapt to new trends and customer behaviors, keeping the service fresh and effective.

Sarah's Vision for the Future

Sarah's strategic implementation of LLMs in TechStyle's ecommerce platform is a testament to the transformative power of AI. By focusing on customer interaction, personalization, and continuous improvement, TechStyle not only meets but exceeds customer expectations. This approach not only boosts sales but also fosters a loyal customer base that feels understood and valued.

Sarah believes in the potential of AI to enhance the human touch rather than replace it. "We're not just using AI to sell products, we're creating an unparalleled shopping experience that combines the efficiency of technology with the warmth of human touch," she says.

Sarah's vision extends beyond immediate sales. She imagines a future where the AI assistant grows with each customer, learning their preferences, and providing even more personalized service over time. This ongoing relationship could transform one-time buyers into lifelong customers, creating a deeply loyal customer base.

Moreover, Sarah is exploring the integration of LLMs with augmented reality (AR) to further enhance the shopping experience. Imagine Shorouk being able to see how the dress looks on her through an AR interface while getting styling tips from the chatbot. This blend of technologies promises to revolutionize online shopping, making it more interactive and immersive.

As TechStyle continues to innovate, Sarah knows that the future of ecommerce lies in the seamless integration of advanced AI with customer-centric strategies, ensuring that every shopping journey is as unique and enjoyable as Shorouk's. By investing in AI technologies and focusing on customer satisfaction, TechStyle sets a benchmark for the industry, showing how businesses can thrive in the digital age.

In conclusion, the application of LLMs in ecommerce at TechStyle is more than just a technological advancement; it's a paradigm shift in how businesses interact with their customers. Through the eyes of Sarah and the experiences of customers like Shorouk, we see a future where AI and human ingenuity work hand in hand to create shopping experiences that are personalized, efficient, and delightfully human.

In the words of Sarah, as she addressed her team at TechStyle: "We're not just building AI systems, we're shaping the future. Let's ensure it's a future we're proud to pass on to the next generation."

And with that, we conclude our exploration of Large Language Models, a journey that has taken us from the basics of natural language processing to the cutting edge of AI research, from the challenges of implementation to the ethical considerations that will shape our AI-driven future. As we move forward, let us carry with us the lessons learned, the questions raised, and the inspiration gained from this chapter in the ongoing story of human innovation.

Conclusion

Large language models (LLMs) are revolutionizing our interaction with the processing of language, demonstrating capabilities that often appear remarkably human-like. These powerful models excel in a wide range of tasks, from text generation and logical reasoning to language translation and sentiment analysis. Their impact is particularly evident in industries like ecommerce and retail, where they enable personalized shopping experiences, AI-driven fashion design, and predictive inventory management.

However, it's crucial to recognize that while LLMs are powerful tools, they require careful oversight and validation. They can potentially reflect biases present in their training data or generate plausible but factually incorrect information. As we continue to develop and deploy these models, we must address critical ethical considerations surrounding privacy, data protection, and responsible use.

In the next chapter, we will discuss AI agents, exploring how these autonomous systems build upon the capabilities of other technologies to interact with their environment and perform tasks with increasing sophistication and autonomy.

CHAPTER 7

AI Agents

Introduction

AI agents are the core unit of Enterprise AI applications. Unlike traditional microservices, agents have the ability to perceive, reason, and act autonomously within their bounded context. At its broad shape, an AI agent is a software program prepared with the principles of artificial intelligence, designed to perceive its environment, process information, and take actions to achieve specific goals. These agents are not inflexible automatons, blindly following a predetermined set of rules, rather, they possess the capacity to adapt, learn, and evolve.

Types of AI Agents

AI agents manifest in diverse forms, each tailored to address specific challenges and requirements. In this section, we will discuss the various kinds of agents we may want to incorporate into our Enterprise AI applications.

Reactive Agents

Reactive agents act as sentinels of the moment and are defined by their ability to perceive and quickly react to their environment. These agents are a mix of simplicity and efficiency, operating on a set of condition–action rules that govern their behavior. Reactive agents observe their surroundings, process sensory data, and then respond with predetermined actions based on the perceived conditions. Reactive agents also have the ability to respond swiftly to changing environments, making reactive agents well-suited for scenarios where timing is critical. Examples of scenarios in which reactive agents would be used are real-time control systems, a robotic arm navigating a complex assembly line, or a video game character reacting to player inputs. For efficiency's sake, these agents lack an internal state (i.e., the ability to maintain memory), a constraint that restricts their capacity for complex reasoning and decision-making.

Logical Agents

Logical agents focus on formal logic and symbolic reasoning, which is different from statistical machine learning. They excel in domains where expert knowledge can be clearly articulated and formalized, such as expert systems, legal reasoning, and theorem proving. These agents work well for clearly defined business processes and can handle the coordination of inputs from their sensor-driven reactive agent cohorts. The challenge of logical agents is the ability to handle incomplete information, a limitation that often arises in dynamic and changing environments. In these cases of uncertainty, higher level agent systems can help in the decision-making process.

Deliberative Agents

While reactive agents are concerned with the immediacy of the present, and logical agents are concerned with the coordination and execution of well-defined business processes, deliberative agents embrace a more contemplative approach that can handle unexpected situations. Also, sometimes called Belief–Desire–Intention (BDI) agents, these agents are designed to not only understand their environment but also maintain an internal state and the ability to reason about their actions. Their desires, represented by a set of goals or objectives, serve as the driving force behind their actions. Their intentions, carefully crafted plans, or strategies, guide their decision-making processes. As they encounter obstacles or unexpected challenges, they do not merely react; instead, they engage in a complex reasoning process, drawing upon their knowledge base, considering the potential consequences, and ultimately formulating a strategic plan to overcome the obstacle and continue their mission. Deliberative agents are designed for more complicated problem-solving and are usually used to coordinate the activities of the logical and reactive agents.

Hybrid Agents

There exists a delicate balance between the time-sensitive reflexes of reactive agents and the contemplative reasoning of deliberative agents. It is within this balance that hybrid agents thrive, harmonizing the duality of these two approaches and harnessing the strengths of each. Hybrid agents possess a layered architecture, where a reactive layer handles time-critical responses with swift efficiency, while a deliberative layer engages in higher-level reasoning and decision-making. Consider a self-driving vehicle whose sensors react to a pedestrian stepping into the road. The reactive layer of the hybrid agent takes over, immediately applying the brakes and averting a potential collision. At the same time, the deliberative layer analyzes the situation, considers alternative routes, and formulates a new plan to reach the destination, all while ensuring the safety of the vehicle and its occupants.

The challenge with hybrid agents is striking the right balance between efficiency and encapsulation. Keeping tasks and functionality separate is a key design goal of software that helps in creating maintainable systems. Small reactive agents can be built as small as possible to ensure the cleanest and most reliable reactions. Adding the additional complexity of deliberation must be balanced with the needs for reliability and encapsulation in your application.

Agent Architecture

Agent architecture requires an intelligent mix of all three types of agents. Large, complex systems that are designed for scale should consider using small reactive agents that feed into logical agents that represent well-defined business processes. Deliberative agents sit at the top of the agent hierarchy and observe the environment in which the other agents operate and make decisions on the behavior of the agent players based on those observed environmental factors. Smaller application teams may opt to build hybrid agents, as simplicity of design and increased responsiveness take precedence over maintainability and scalability considerations.

A major design consideration in designing agents, however, is the documentation and observability of explicit incentive structures. While it may be faster and cheaper to build hybrid agents that both react to sensor input and perform complicated reasoning based on that input, as the agents become more complex over time, it is difficult to parse out the reasons for certain decisions. This problem of "Black Box AI" has been seen in complicated neural networks, in which there is no clear understanding of why the AI acted the way it did, impacting root cause analysis situations in the inevitable cases of failure. This is the reason why we strongly advocate for a tiered agent architecture that divides action and reasoning duties at multiple levels among independently implemented AI agents.

Agent Communication and Coordination

Most of us have experienced AI in the form of chatbots like ChatGPT or expert systems such as Watson or chess engines. AI agents are unique in their ability to communicate, coordinate, and collaborate with one another and are reshaping the way we think about enterprise application design.

AI agents harmonize their actions, exchange information, and leverage each other's strengths to solve complex problems and achieve shared goals that would be insurmountable for any single entity alone. Many of these coordination patterns have been utilized in machine learning design such as convolutional neural networks, and Enterprise AI applications utilize these patterns on the much higher level of business processes.

Agent Communication

At the heart of agent communication lies a sophisticated language that exchanges information, conveys intentions, and coordinates their actions with precision and clarity. This language, known as the Agent Communication Language (ACL), is a structured format that defines the syntax and semantics of the messages exchanged between agents, ensuring that they can communicate effectively, regardless of their origins or the platforms upon which they were developed. One of the most widely used ACLs is the FIPA-ACL (Foundation for Intelligent Physical Agents–Agent Communication Language), a standardized language developed by the IEEE Computer Society. This language defines a set of performatives, or speech acts, that agents can use to convey different types of messages, such as requests, queries, or assertions, enabling them to engage in meaningful and coherent conversations. Consider a scenario where two AI agents, each with unique capabilities and specialized knowledge, are tasked with solving a complex logistics problem. Through the exchange of messages using the ACL, these agents can share information about their respective domains, negotiate strategies, and coordinate their efforts to develop an optimal solution that leverages their collective intelligence.

Agent Communication Protocols

While the ACL provides the language for agent communication, it is the agent communication protocols that define the rules and conventions that govern the exchange of messages, ensuring that the agents engage in meaningful and coherent conversations. These protocols address issues such as turn taking, message sequencing, and error handling, much like the conductor of an orchestra ensures that each musician plays their part at the right time and in harmony with the rest of the ensemble. One such protocol is the Contract Net Protocol (CNP), which is designed for task allocation and coordination among agents. In this protocol, an agent assumes the role of a manager and announces a task or problem to be solved. Other agents can then evaluate their capabilities and submit bids or proposals to take on the task or contribute to the solution. The manager agent (a deliberator) then evaluates the proposals and assigns the task or tasks to the most suitable agents, facilitating effective coordination and resource allocation.

By communicating and coordinating their actions through agent communication languages and protocols, agents can collectively optimize the entire supply chain, identifying bottlenecks, streamlining processes, and minimizing costs and inefficiencies. The power of multi-agent systems lies in their ability to distribute tasks, leverage parallel processing, and adapt to dynamic environments. As individual agents learn and evolve, the collective intelligence of the system grows, enabling it to tackle increasingly complex challenges and adapt to changing circumstances with agility and resilience.

Agent Coordination Mechanisms

When you think of agent communication and coordination, the choice of coordination mechanism plays an important role in ensuring that the agents work together harmoniously. These coordination mechanisms determine how agents interact and collaborate to achieve their goals, resolving conflicts and ensuring that their actions align with the overall objectives of the system. One such coordination mechanism is centralized coordination, where a central coordinator agent is responsible for managing and coordinating the activities of other agents in the system. This approach is assigning tasks, resolving conflicts, and ensuring that the overall system goals are met.

Another mechanism is distributed coordination, where there is no central authority. Instead, agents communicate and negotiate with each other directly to coordinate their actions and resolve conflicts. This approach is more flexible and scalable but can be more complex to implement.

Organizational coordination is another mechanism that involves defining organizational structures, roles, and norms that govern the behavior and interactions of agents within the system. Agents are assigned specific roles and responsibilities, and their actions are guided by predefined norms or rules.

Agent Development Tools and Frameworks

A diverse array of tools and frameworks have emerged, each with its unique strengths and specializations. One such tool is JADE (Java Agent DEvelopment Framework), a widely used open-source framework for developing multi-agent systems in compliance with FIPA standards. JADE provides a runtime environment, a set of graphical tools, and a library of reusable components, empowering developers to create and deploy agent-based applications with ease and efficiency. Another powerful instrument in the agent developer's arsenal is JACK (Java Agent Compiler & Kernel), a commercial agent development environment based on the Belief–Desire–Intention (BDI) architecture. JACK offers a graphical environment for designing and implementing BDI agents, as well as tools for debugging and deploying agent-based systems, making it an ideal choice for developers working in domains that require sophisticated reasoning and decision-making capabilities.

For those who use the open-source philosophy, JASON (Java-based interpreter for an extended version of AgentSpeak) stands as a beacon of innovation. This open-source interpreter for an extension of the AgentSpeak language, based on the BDI architecture, provides a platform for developing and running multi-agent systems, complete with tools for debugging, visualization, and integration with external environments.

In modern application development, where the need for agility, scalability, and seamless integration is paramount, developers have embraced the power of open APIs and cloud-based services. The OpenAPI specification, formerly known as Swagger, has emerged as an industry standard for defining and documenting APIs, enabling developers to create, consume, and integrate APIs with ease. By leveraging tools like the OpenAPI Generator, developers can generate client libraries, server stubs, and documentation for APIs defined using the OpenAPI specification, streamlining the development process, and facilitating seamless communication between AI agents and other services.

Amazon Web Services (AWS) offers a comprehensive suite of services and tools, including Amazon Bedrock, a modern application development framework that simplifies the process of building, deploying, and managing AI applications on AWS. By leveraging Amazon Bedrock, developers can create AI agents that seamlessly integrate with other AWS services, such as Amazon Lex for conversational interfaces, Amazon Rekognition for image and video analysis, and Amazon SageMaker for building, training, and deploying machine learning models. This powerful combination of open APIs, cloud services, and agent development tools empowers developers to create intelligent systems that can perceive, reason, and act across a wide range of domains, from natural language processing and computer vision to predictive analytics and decision support.

Agent Development Methodologies

The creation of AI agents requires more than just the right tools; it demands a methodology that guides the development process from conception to deployment. Several methodologies have emerged, each offering a unique approach to the art of agent development. The Prometheus methodology is a comprehensive approach that covers the entire development life cycle, from requirements analysis and system specification to architectural design and detailed implementation. It provides a structured framework for identifying agents, their roles, interactions, and organizational structures, ensuring that the resulting agent-based system is cohesive, scalable, and aligned with the underlying business objectives.

For those who utilize the principles of goal-oriented requirements engineering, the Tropos methodology offers a compelling approach. By integrating concepts from agent-oriented software engineering and goal-oriented requirements analysis, Tropos guides developers through the process of identifying stakeholder goals, translating them into system requirements, and ultimately designing and implementing agent-based solutions that align with these goals. In agile development, where iterative and incremental approaches reign supreme, methodologies such as Scrum and Extreme Programming (XP) have found their place in the agent development landscape. These methodologies emphasize collaboration, continuous delivery, and rapid adaptation, enabling development teams to respond quickly to changing requirements and evolving business needs.

Best Practices in Agent Development

One of the most critical best practices in agent development is the adherence to the principles of modularity and separation of concerns. AI agents should be designed with clear boundaries and well-defined responsibilities. This approach not only enhances the maintainability and scalability of agent-based systems but also facilitates collaboration among development teams.

Another essential best practice is the emphasis on testability and verification. As AI agents become increasingly complex and autonomous, ensuring their reliability and predictability becomes paramount. By considering methodologies such as test-driven development and continuous integration, developers can catch defects early in the development cycle, minimizing the risk of costly errors and ensuring that the agents they create are truly fit for purpose.

Developers must also consider best practices for cloud native application development, such as containerization, microservices architecture, and Infrastructure as Code (IaC). By leveraging tools like Docker and Kubernetes, developers can package and deploy their AI agents as lightweight, portable containers, ensuring consistent behavior across different environments and enabling seamless scaling and orchestration.

Furthermore, the integration of open APIs and cloud services like AWS or GCP or Azure into agent-based systems demands a deep understanding of API design principles and best practices for cloud service integration. Developers must carefully consider factors such as API versioning, security, and performance, while also leveraging cloud native patterns and practices to ensure the scalability, reliability, and cost effectiveness of their solutions.

AI Agent Use Cases

There are several areas today in which AI agents are integrated into corporate enterprise portfolios. In this section, we will provide several examples of AI agent applications that we have seen in the enterprise.

Customer Service and Ecommerce

The first and most popular use of AI agents takes the form of customer service chatbots. These bots are implemented on most corporate consumer sites today and range from simple bots with a hard coded script to personalized bots that remember user preferences and conversational history. The move from chat to voice recognition was obvious, and we anticipate that most call centers will implement some version of voice-enabled AI agents in their call centers within the next couple years.

AI has been used in the form of recommender systems in ecommerce since the mid-2000s, and we are now seeing AI agents harness this capability in their ecommerce customer service bots, providing interactive and customized recommendations for ecommerce customers. From agents that help you choose clothing based on your personal style to agents that package industrial purchases based on specific business needs, many companies are already experimenting with these personalized agents.

Decision Support Systems

AI agents implemented as part of decision support systems can be extremely useful at sifting through large volumes of data to identify anomalies and emerging patterns. When fed quality data and provided with explicit and well-understood incentive structures, decision support agents offer a level of analysis that is impossible for human analysts alone.

While AI tools have been used in data analysis for many years, AI agents can provide a level of configurability and interaction to the executive that has up to now been limited to the purview of trained data engineers.

Intelligent Tutoring Systems

Some cutting-edge schools have introduced tutoring agents to their elementary students as a means of personalized education, identifying areas of struggle, and tailoring their approach to ensure that concepts and lessons resonate with clarity and understanding. Through adaptive algorithms and intelligent user interfaces, AI agents in intelligent tutoring systems can create immersive and engaging learning experiences, assisting teachers and providing additional learning resources to students and parents.

Business Process Automation

From data entry and document processing to workflow management, the reasoning capabilities of AI agents allow businesses to move beyond mere task automation and move to more complex business process automation. BPA in the form of rigid rules-based engines and complex expert systems has been difficult to implement in business critical environments. The learning capacity of AI agents allows for these processes to be more flexible and easier to maintain than previous monolithic business rule engines. These systems can provide reliable decision-making capabilities as well as optimize the efficiency and productivity of human staff by automating complex, repetitive, and error-prone tasks.

IT Operations Automation

Observability and anomaly detection alerting systems form the backbone of IT operations. Overworked systems operations support staff are some of the most enthusiastic users of AI agent applications as they relieve the stress of having to keep all eyes peeled on every change event and anomaly across the enterprise all by themselves. AIOp agents coordinate together to form a holistic picture of enterprise operations help, remediating common events such as memory overloads and system restarts. SecOps agents also work to automatically apply the latest patches and constantly scan for the newest security vulnerabilities.

Robotic Process Automation (RPA)

Robotic Process Automation (RPA) agents mimic and amplify physical human actions and interactions, automating repetitive tasks across various software platforms. With each keystroke and mouse click, the agent replicates the actions of its human counterparts, extracting data, consolidating information, and generating reports with a speed and precision that transcends the limitations of manual processes. Several RPA tools exist on the market today, and the coordination of AI agents to manage and conduct RPA agents promises new levels of business productivity.

Intelligent Automation

AI agents possess the ability to not only automate repetitive tasks but also to engage in complex cognitive processes and decision-making. AI agents superpass the limitations of traditional automation systems as they are capable of analyzing unstructured data, detecting patterns, and making informed decisions that once belonged solely to the domain of human expertise.

From streamlining business processes and elevating customer experiences, to safeguarding digital infrastructures, AI agents are the driving force behind a transformation that promises to reshape the modern enterprise.

Conclusion

In this chapter, we have continued our discussion of Enterprise AI application technical design with a focus on AI agents. We have discussed the different types of agents and the best practices for designing AI applications that consist of the collaboration of several AI agents working in tandem. We concluded this section with an overview of several use cases in which we see AI agents playing valuable roles in the enterprise.

In the next two chapters, we will be discussing the critical elements of testing and test automation in building your Enterprise AI application.

PART III

Maintaining Your AI Application

CHAPTER 8

Testing Your Enterprise AI Application

Introduction

Machine learning models and intelligent LLM systems are rapidly transforming industries, and the importance of robust testing cannot be overstated. As enterprises across various sectors use AI technologies, ensuring the reliability, accuracy, and ethical integrity of these systems becomes increasingly important. Testing, which has long been a cornerstone of traditional software development, takes on a new level of significance when applied to the world of Enterprise AI.

AI-powered systems are responsible for making critical decisions that impact millions of lives, from diagnosing medical conditions and approving loans to optimizing supply chains and managing energy grids. In such use cases, the consequences of failure could be catastrophic. A single erroneous prediction or incorrect decision could lead to financial losses, legal liabilities, and even endanger human lives. This stark reality underscores the importance of rigorous testing in Enterprise AI

A. Cagle and A. M. C. Ahmed, *Architecting Enterprise AI Applications*,
https://doi.org/10.1007/979-8-8688-0902-6_8

development. Beyond the obvious risks of system failures, testing also plays a significant role in addressing the unique challenges posed by AI systems. Unlike traditional software, which operates on predefined rules and logic, AI models are continually learning based on incoming information and make decisions based on complex patterns and relationships. This inherent complexity introduces new sources of potential errors, such as biased or incomplete training data, overfitting or underfitting of models.

Enterprise AI systems often operate in dynamic and changing environments, where conditions can shift rapidly. A model trained on historical data may become outdated or biased as new patterns emerge, leading to inaccurate predictions or unfair decisions. Continuous testing and monitoring are essential to ensure that these systems remain reliable, accurate, and aligned with the evolving needs of the enterprise. Consider the example of FinTech Corporation, a leading provider of AI-powered financial services. FinTech's machine learning models are responsible for assessing loan applications, evaluating creditworthiness, and detecting fraudulent activities. Without rigorous testing, a flaw in the model's decision-making process could lead to unfair loan denials, exposing the company to legal risks and reputational damage. Conversely, a well-tested and validated model can help FinTech maintain a competitive edge by providing accurate risk assessments, streamlining processes, and improving customer experiences.

Recognizing the critical importance of testing in Enterprise AI applications, organizations must adopt a comprehensive testing strategy that addresses the unique challenges posed by these systems. This strategy should encompass various testing methodologies, from traditional software testing practices to AI-specific techniques, ensuring end-to-end validation of the entire AI pipeline. At the foundation of this strategy lies unit testing, which focuses on validating individual components or functions within the AI system. This includes testing data preprocessing routines, model training and evaluation processes, and inference logic.

146

Effective unit testing not only ensures the correctness of individual components but also facilitates easier debugging and maintenance of the overall system. As AI systems become more complex, involving multiple interconnected components, integration testing becomes increasingly important. This type of testing verifies that the various components of the AI pipeline, from information ingestion to model deployment, seamlessly integrate and function as expected when combined. By simulating real-world scenarios and edge cases, integration testing helps identify and resolve potential conflicts or compatibility issues before deployment.

System testing takes a holistic approach, evaluating the AI system as a complete end-to-end solution within the enterprise environment. This includes assessing performance and scalability, ensuring that the system can handle the anticipated workloads and data volumes without compromising accuracy or efficiency. Load testing, stress testing, and other performance-related tests are essential components of system testing for Enterprise AI. Beyond traditional testing methodologies, Enterprise AI introduces unique challenges that require specialized testing strategies. Model validation techniques, such as confusion matrix analysis and metrics like accuracy, precision, and recall, are vital for evaluating the performance and reliability of machine learning models. Bias and fairness testing, which involves identifying and mitigating potential sources of bias in training data and model decisions, is essential for ensuring ethical and equitable outcomes. Robustness testing is another critical aspect of Enterprise AI testing, focusing on the system's ability to handle adversarial inputs, edge cases, and unexpected conditions. This includes simulating various types of attacks or corruptions to assess the model's resilience and identify potential vulnerabilities. As Enterprise AI systems continue to evolve and become more pervasive, the need for comprehensive testing strategies will only grow. By adopting a rigorous and holistic approach to testing, organizations can ensure the reliability, accuracy, and ethical integrity of their AI solutions, mitigating risks and fostering trust among stakeholders and end-users alike.

Types of Testing

In this section, we explore the various types of testing for developing robust and reliable Enterprise AI systems. Understanding and implementing these testing strategies ensure the accuracy, performance, and reliability of AI applications within enterprise environments.

Unit Testing for Enterprise AI

Unit testing is the foundation of any robust testing strategy, and its importance is magnified in the context of Enterprise AI. At its core, unit testing involves testing individual units or components of a software system to verify their correctness and functionality. In the case of Enterprise AI, these units can range from data preprocessing functions and feature engineering pipelines to model training and inference logic. The significance of unit testing in Enterprise AI development cannot be overstated. Machine learning models and AI systems are inherently complex, with numerous interdependent components that must work seamlessly together. A single bug or error in any of these components can have cascading effects, leading to inaccurate predictions, biased decisions, or even complete system failures. Imagine a scenario at TechCorp, a leading technology company that develops AI-powered solutions for various industries. TechCorp's data science team is working on a machine learning model for predictive maintenance in manufacturing facilities. Without proper unit testing, a seemingly trivial error in the data preprocessing code could result in corrupted or missing data being fed into the model. This, in turn, could lead to incorrect predictions, potentially causing costly equipment failures or compromising worker safety. By implementing rigorous unit testing practices, TechCorp's developers can catch and fix such errors early in the development cycle, ensuring the reliability and accuracy of individual components before integrating them into the larger system.

When it comes to unit testing in Enterprise AI, there are several techniques and best practices that developers should adopt to ensure thorough and effective testing. One common approach is to create test suites that cover a wide range of scenarios and edge cases for each unit or component. These test suites should include both positive and negative test cases, verifying that the component behaves correctly under expected conditions and handles errors or invalid inputs gracefully. For example, when testing a data preprocessing function that handles missing values in a dataset, the test suite should include cases with complete data, data with missing values in various columns or patterns, and edge cases such as datasets with all values missing or entirely empty datasets. Another essential technique in Enterprise AI unit testing is the use of mocking and stubbing. Machine learning models and AI systems often rely on external dependencies, such as databases, APIs, or other third-party services. Mocking and stubbing allow developers to isolate the component under test by simulating these dependencies, ensuring that tests are focused, deterministic, and independent of external factors.

Best practices in Enterprise AI unit testing also emphasize code quality, maintainability, and collaboration. Developers should strive for clean, modular code that adheres to established coding standards and follows the principles of testability, such as separation of concerns and loose coupling. Continuous integration and code review practices can further enhance the quality and reliability of the codebase.

Integration Testing in Enterprise Environments

While unit testing is important for validating individual components, Enterprise AI applications are complex ecosystems comprising multiple interconnected components. Integration testing focuses on verifying that these components work together seamlessly, ensuring end-to-end functionality and identifying potential compatibility issues or conflicts. Integration testing becomes particularly important due to the diverse range of components involved in the AI pipeline. From data ingestion and preprocessing to model training, deployment, and inference, each stage of the pipeline may involve different technologies, frameworks, and tools. Consider the example of FinTech Corporation, a leading provider of AI-powered financial services. FinTech's loan approval system consists of several components, including a data ingestion module that collects and preprocesses customer information, a machine learning model that evaluates creditworthiness, and a decision engine that applies business rules and regulations to the model's predictions. Without proper integration testing, it is possible for these components to work correctly in isolation but fail when combined. For instance, a change in the data format or structure between the data ingestion module and the model training pipeline could lead to incorrect or incomplete data being fed into the model, resulting in inaccurate predictions and potentially unfair loan decisions.

Integration testing in Enterprise AI environments often reveals a range of common issues that can arise due to the complexity and diversity of the components involved. One of the most prevalent challenges is data compatibility and consistency. Machine learning models and AI systems are heavily dependent on underlying data processes, and any inconsistencies or mismatches in data formats, structures, or semantics can lead to integration failures. To address this issue, organizations should implement robust data governance practices, standardizing data formats, and establishing clear data pipelines. Additionally, data validation and transformation routines should be integrated into the testing process to ensure that data remains consistent and compatible across different components.

Another common challenge in Enterprise AI integration testing is the management of dependencies and versioning. As AI systems evolve and new components or libraries are introduced, ensuring compatibility and tracking version changes become critical. Automated dependency management tools and strict version control practices can help mitigate these issues. Organizations should also consider implementing continuous integration and deployment pipelines specifically tailored for AI systems. These pipelines can automate the process of building, testing, and deploying AI components, ensuring that integration issues are detected and addressed early in the development cycle.

System Testing for Enterprise AI Applications

While unit and integration testing are essential for validating individual components and their interactions, system testing takes a holistic approach to evaluating the entire Enterprise AI solution within its intended deployment environment. This type of testing aims to verify that the AI system meets the specified requirements, performs as expected, and delivers the desired business value. System testing becomes particularly needed due to the potential impact of these systems on critical business operations and decision-making processes. The failure or underperformance of an AI system can have far-reaching consequences, ranging from financial losses and operational inefficiencies to legal liabilities and reputational damage. Consider the example of a large retail corporation that has implemented an AI-powered inventory management system. System testing would involve evaluating the entire solution, including the data ingestion processes, machine learning models for demand forecasting and supply chain optimization, and the integration with existing enterprise systems such as inventory management and ordering platforms. During system testing, various scenarios and use cases would be simulated to ensure that the AI system can handle real-world complexities and edge cases. This may involve testing with historical data, simulating seasonal demand fluctuations, or introducing unexpected events such as supply chain disruptions or changes in customer preferences.

In addition to functional testing, system testing must also consider performance and scalability aspects. Machine learning models and AI systems are often computationally intensive and require significant resources, such as CPU, GPU, and memory. As these systems are deployed in production environments, it is essential to ensure that they can handle the anticipated workloads and data volumes without compromising accuracy or efficiency. Performance testing involves evaluating various metrics, including latency (the time it takes for the system to respond to a request), throughput (the number of requests the system can handle concurrently), and resource utilization (such as CPU and memory consumption).

Scalability testing, on the other hand, focuses on the ability of the AI system to scale up or down to meet changing demand or workload requirements. This may involve testing the system under different load conditions, simulating scenarios with varying numbers of concurrent users or data volumes, and assessing the system's ability to scale horizontally (by adding more computational resources) or vertically (by increasing the capacity of existing resources). Organizations should also consider implementing monitoring and alerting mechanisms to continuously evaluate the performance and scalability of their Enterprise AI systems in production environments. This allows for proactive identification and resolution of performance bottlenecks, ensuring that the AI solution remains efficient and reliable even as usage patterns or data volumes evolve over time.

Enterprise AI-Specific Testing Considerations

While traditional software testing methodologies provide a solid foundation for validating Enterprise AI applications, the unique nature of these solutions necessitates additional testing considerations tailored specifically for AI models and algorithms. From model validation and bias mitigation to robustness testing and adversarial input handling, these AI-specific testing practices are important for ensuring the reliability, fairness, and trustworthiness of AI solutions.

Model Validation in Enterprise AI

Model validation is a critical step in the development of Enterprise AI systems, ensuring that the models perform as expected and meet the desired accuracy, precision, and recall metrics. This process involves evaluating the model's performance using various validation techniques to ensure its reliability and effectiveness in real-world applications.

Accuracy, Precision, and Recall

At the core of any Enterprise AI system lies one or more machine learning models responsible for making predictions, classifications, or recommendations based on input data. Model validation often involves evaluating metrics such as accuracy, precision, and recall. Accuracy measures the overall correctness of the model's predictions, calculated as the ratio of correct predictions to the total number of predictions made. While accuracy is a useful metric, it can be misleading in cases where the dataset is imbalanced or skewed toward a particular class. Precision and recall provide a more nuanced view of the model's performance by considering the trade-off between false positives and false negatives. Precision measures the proportion of true positive predictions among all positive predictions made by the model, while recall measures the proportion of true positive predictions among all actual positive instances in the data.

For example, consider a financial institution that has developed an AI-powered fraud detection system. In this scenario, precision would indicate the system's ability to correctly identify genuine fraudulent transactions, while recall would measure its ability to capture all instances of fraud within the dataset.

Confusion Matrix Analysis

To gain deeper insights into a model's performance and identify areas for improvement, Enterprise AI developers often employ confusion matrix analysis. A confusion matrix is a tabular representation that compares the model's predicted output against the actual ground truth, providing a comprehensive view of the model's performance across different classes or categories. By analyzing the confusion matrix, developers can identify specific patterns of misclassification or errors made by the model. This information can be invaluable for understanding the model's strengths and weaknesses, as well as for identifying potential sources of bias or data quality issues that may be impacting its performance. For instance, in a computer vision application for defect detection in manufacturing, a confusion matrix could reveal that the model is consistently misclassifying certain types of defects or struggling to differentiate between similar defect patterns. Armed with this information, developers can take targeted actions to improve the model's performance, such as collecting more diverse training data, fine-tuning the model architecture, or adjusting the hyperparameters.

Bias and Fairness Testing in Enterprise AI

Ensuring fairness and mitigating bias are paramount in Enterprise AI development. Bias and fairness testing involves identifying and addressing any biases in the AI models to ensure that the solutions are equitable and do not unfairly discriminate against any group. This section explores techniques and metrics used to evaluate and enhance the fairness of AI systems.

Identifying and Mitigating Bias

One of the most significant challenges in Enterprise AI application development is ensuring the fairness and impartiality of models. Bias, whether intentional or unintentional, can creep into AI systems through various sources, including biased training data, algorithmic biases, or biases introduced during data preprocessing and feature engineering. Identifying and mitigating these biases are vital to maintain the integrity and trustworthiness of solutions, particularly in domains where decisions can have far-reaching impacts on individuals or communities, such as recruitment, lending, or health care. Bias testing in Enterprise AI involves a range of techniques and methodologies. One approach is to analyze the training data used to build the model, looking for potential sources of bias or underrepresentation of certain groups or characteristics. Statistical techniques, such as measuring the distribution of sensitive attributes (e.g., gender, race, and age) within the dataset, can help identify potential data biases. Another avenue for bias testing is to examine the model's outputs and decisions across different subgroups or sensitive attributes. This can be achieved through techniques like disparate impact analysis, which compares the model's performance across different groups and identifies potential disparities or discriminatory patterns.

Imagine a scenario where a large retail corporation has developed an AI-powered system for targeted marketing and personalized product recommendations. During bias testing, the developers discover that the model is consistently underrepresenting certain demographic groups in its recommendations, potentially due to biases in the training data or the features used for personalization. By identifying and addressing these biases, the company can ensure that its marketing efforts are fair and inclusive, ultimately improving customer satisfaction and avoiding potential legal or reputational risks.

Fairness Metrics

In addition to identifying sources of bias, Enterprise AI developers often employ fairness metrics to quantify and measure the level of bias present in a model's predictions or decisions. These metrics provide a quantitative framework for evaluating and comparing the fairness of different models or algorithms, enabling developers to make informed decisions about which models to deploy or which techniques to employ for bias mitigation. One commonly used fairness metric is demographic parity, which measures the difference in the likelihood of a positive outcome (e.g., loan approval, and job offer) between different demographic groups. Another metric, equal opportunity, focuses on ensuring that individuals with similar qualifications or characteristics have an equal chance of receiving a positive outcome, regardless of their sensitive attributes. By incorporating fairness metrics into their testing and evaluation processes, Enterprise AI teams can establish quantitative benchmarks for acceptable levels of bias and continuously monitor their models for compliance with these benchmarks, ensuring that the solutions they deploy meet the highest standards of fairness and ethical integrity.

Robustness Testing for Enterprise AI

Robustness testing assesses the resilience of AI systems against adversarial inputs and other challenging conditions. It ensures that the models can handle unexpected variations and maintain performance under stress. This section covers strategies for testing and improving the robustness of AI models to ensure their reliability in real-world scenarios.

Handling Adversarial Inputs

In the real world, Enterprise AI systems are often exposed to a wide range of input data, including adversarial inputs designed to intentionally mislead or deceive the model. Robustness testing focuses on evaluating the resilience of these systems against such adversarial inputs, as well as other forms of noise, corruption, or unexpected variations in the input data. Adversarial inputs can take many forms, from carefully crafted images or text designed to fool computer vision or natural language processing models to malicious attempts to inject biased or corrupted data into the system. These inputs can potentially cause the AI model to make incorrect predictions, fail to detect critical patterns, or even exhibit unexpected and potentially harmful behaviors. To mitigate these risks, Enterprise AI developers employ various techniques for robustness testing, including adversarial attack simulations, input perturbation, and data augmentation. Adversarial attack simulations involve generating adversarial inputs using specialized algorithms or tools and testing the model's performance and behavior under these conditions. Input perturbation involves introducing controlled noise or variations to the input data and observing the model's response, helping to identify potential vulnerabilities or edge cases that may not be adequately addressed by the training data alone. Data augmentation, on the other hand, involves expanding the training dataset with synthetic or modified samples, increasing the model's exposure to a broader range of input variations and improving its robustness to real-world scenarios.

Stress Testing

In addition to handling adversarial inputs, Enterprise AI systems must also be capable of performing reliably under extreme conditions or high-stress situations. Stress testing is a component of robustness testing, focused on evaluating the system's performance and stability when subjected to extreme workloads, resource constraints, or other challenging conditions. For example, in a manufacturing setting, an AI-powered quality control system may need to process and analyze a large volume of high-resolution images or sensor data in real time, while maintaining strict accuracy and latency requirements. Stress testing would involve simulating scenarios

157

with varying workloads, resource constraints (e.g., limited CPU or memory), or other challenging conditions to ensure that the system can maintain its performance and reliability under these circumstances. Similarly, in the financial sector, an AI-powered fraud detection system may need to process millions of transactions concurrently, while adhering to strict regulatory requirements and security protocols. Stress testing would involve simulating high-volume transaction scenarios, network congestion, or other adverse conditions to validate the system's ability to maintain its accuracy and responsiveness even under extreme loads.

Tools and Frameworks for Enterprise AI Testing

As the adoption of Enterprise AI continues to grow, so too does the need for robust tools and frameworks that can support the testing and validation of these complex systems. From unit testing frameworks to specialized libraries for model validation and fairness analysis, the landscape of Enterprise AI testing tools is rapidly evolving to meet the unique challenges posed by these cutting-edge technologies.

Popular Tools for Enterprise AI Testing

In this section, we will introduce a number of popular and industry-leading tools that are used for testing Enterprise AI applications.

PyTest

PyTest is a widely popular testing framework for Python, the programming language of choice for many data scientists and AI developers. While not specifically designed for AI testing, PyTest's flexibility and extensive ecosystem of plugins make it a powerful tool for validating Enterprise AI systems. One of the key advantages of PyTest is its simplicity and ease of use. With a clean and intuitive syntax, developers can quickly write and execute unit tests, as well as more complex integration and functional tests. PyTest also supports powerful features like parallelization, which can significantly speed up test execution times, and fixtures, which allow for efficient setup and teardown of test environments. PyTest can be used to test various components of the AI pipeline, from data preprocessing and feature engineering to model training and inference logic. For example, developers can write unit tests to ensure that their data cleaning routines are correctly handling missing values or outliers, or that their feature scaling and encoding functions are producing the expected output. Additionally, PyTest's plugin ecosystem offers extensions specifically tailored for AI testing. The pytest-ml plugin, for instance, provides utilities for testing machine learning models, including tools for generating and validating test data, as well as utilities for evaluating model performance using metrics like accuracy, precision, and recall.

TensorFlow Test

For organizations heavily invested in the TensorFlow ecosystem, the TensorFlow Test library provides a comprehensive set of tools for testing and debugging TensorFlow-based models and applications. As an integral part of the TensorFlow ecosystem, It offers seamless integration with other components, ensuring a consistent and streamlined testing experience. One of the key features of TensorFlow Test is its support for dynamic computation graphs, which are central to the TensorFlow architecture. This allows developers to write tests that accurately simulate the behavior of their models, ensuring that the computed outputs match the expected results under various input conditions. It also includes utilities for testing and debugging distributed TensorFlow systems, which are becoming increasingly common in Enterprise AI deployments where scalability and performance are critical. Developers can use these tools to simulate distributed training and inference scenarios, identify potential bottlenecks or synchronization issues, and validate the correctness of their models in a distributed environment.

IBM AI Fairness 360

As concerns around bias and fairness in AI systems continue to grow, tools like IBM AI Fairness 360 have emerged to help organizations ensure the ethical and responsible development of their Enterprise AI solutions. IBM AI Fairness 360 is an open-source toolkit that provides a comprehensive set of metrics and algorithms for detecting and mitigating bias in machine learning models. One of the standout features of AI Fairness 360 is its support for a wide range of bias mitigation algorithms, including preprocessing techniques for reweighting or transforming training data, in-processing techniques for modifying the model's optimization objective, and postprocessing techniques for adjusting the model's predictions. The toolkit also includes a rich set of fairness metrics, such as demographic parity, equal opportunity, and disparate impact, which can be used to quantify and compare the fairness of different models or algorithms. These metrics are particularly valuable in Enterprise AI scenarios where decision-making processes can have significant impacts on individuals or groups, such as in hiring, lending, or healthcare applications.

Frameworks and Libraries

In addition to dedicated testing tools, Enterprise AI developers also rely on a range of frameworks and libraries that provide essential functionality for building, training, and evaluating machine learning models. These frameworks often include built-in testing utilities or integrate seamlessly with popular testing tools, making them invaluable resources for Enterprise AI testing.

Scikit-Learn

Scikit-learn is a widely used machine learning library for Python, renowned for its simplicity, consistency, and extensive documentation. While not specifically designed for Enterprise AI applications, scikit-learn provides a solid foundation for building and testing a wide range of machine learning models, from classical algorithms like linear regression and decision trees to more advanced techniques like support vector machines and ensemble methods. One of the key strengths of scikit-learn for Enterprise AI testing is its emphasis on modularity and code reusability. By breaking down complex machine learning pipelines into modular components, such as data transformers, estimators, and model evaluation metrics, scikit-learn allows developers to test each component independently, ensuring that errors or issues can be isolated and addressed more efficiently.

Also, scikit-learn includes utilities for model evaluation and validation, such as cross-validation techniques, scoring functions, and tools for visualizing model performance. These features can be invaluable for Enterprise AI teams looking to rigorously test and validate their models before deploying them in production environments.

Keras

Keras is a high-level neural network API, designed to enable fast experimentation and prototyping of deep learning models. While Keras can be used as a stand-alone library, it is often used in conjunction with lower-level frameworks like TensorFlow or PyTorch, which handle the computationally intensive aspects of model training and inference. For Enterprise AI testing, Keras offers several advantages. Its user-friendly syntax and modular architecture make it easier to write and maintain unit tests for individual components of deep learning models, such as layers, activation functions, and loss functions. Additionally, Keras includes built-in support for callbacks, which can be used to monitor and validate model performance during training, enabling more effective testing and debugging. It also integrates seamlessly with popular testing frameworks like PyTest, allowing developers to leverage the full suite of PyTest's features, such as fixtures and parametrization, when testing their Keras-based models.

The Keras ecosystem includes several third-party libraries and extensions that can further enhance Enterprise AI testing capabilities. For example, the Keras Tuner library provides tools for hyperparameter tuning and model selection, which can be invaluable for optimizing model performance and ensuring that the deployed models are as accurate and efficient as possible.

Additional Tools

One of the key challenges in Enterprise AI testing is the inherent complexity and opaqueness of many machine learning models, particularly deep neural networks. These models often behave as "black boxes," making it difficult to understand and interpret their decision-making processes. This opacity can pose significant challenges when it comes to testing and debugging, as traditional techniques for tracing code execution and identifying root causes may not be as effective.

To address this challenge, Enterprise AI teams are increasingly turning to techniques like interpretable machine learning and explainable AI (XAI). These approaches aim to shed light on the inner workings of complex models, providing insights into the factors that influence their predictions or decisions. By leveraging tools like SHAP (SHapley Additive exPlanations) or LIME (Local Interpretable Model-agnostic Explanations), developers can gain a better understanding of how their models are behaving, allowing them to identify potential biases, anomalies, or areas for improvement more effectively.

Another key consideration in Enterprise AI testing is the need for continuous monitoring and validation. Different from traditional software systems, which typically operate under static conditions, machine learning models are often deployed in dynamic environments where input data, usage patterns, and external factors can change over time. This dynamism introduces the risk of model drift, where a model's performance gradually degrades as the underlying data distribution or real-world conditions shift.

To mitigate this risk, Enterprise AI teams are adopting practices like continuous monitoring and automated model retraining. By continuously collecting and analyzing data from their deployed models, organizations can detect signs of model drift or performance degradation early, triggering processes for model re-evaluation, retraining, or even replacement.

Tools like TensorFlow Extended (TFX) and Kubeflow provide end-to-end machine learning pipelines that support continuous monitoring, validation, and automated model retraining, enabling organizations to maintain the reliability and accuracy of their Enterprise AI systems over time.

Moreover, as Enterprise AI systems become more pervasive and impactful, the importance of testing for fairness, bias, and ethical considerations cannot be overstated. Tools like IBM AI Fairness 360 and Google's What-If Tool provide organizations with the means to evaluate their models for potential biases and discrimination, ensuring that their AI systems are not perpetuating or amplifying societal inequities.

As we look to the future of Enterprise AI testing, it's clear that the challenges will only continue to grow in complexity. The emergence of new AI paradigms, such as federated learning, multi-task learning, and few-shot learning, will likely introduce novel testing requirements and considerations. Additionally, the increasing adoption of AI systems in high-stakes domains like health care, finance, and critical infrastructure will raise the bar for testing and validation, demanding even higher levels of rigor, transparency, and assurance.

To meet these challenges, the Enterprise AI testing ecosystem will need to continue evolving, with new tools, frameworks, and methodologies emerging to address the unique needs of these cutting-edge technologies. Collaboration between academia, industry, and open-source communities will be essential in driving this evolution, fostering the development of innovative testing approaches and best practices. Ultimately, the success of Enterprise AI will hinge on our ability to build and deploy these systems with confidence and trust. By embracing a culture of rigorous testing and validation, supported by robust tools and frameworks, organizations can unlock the transformative potential of AI while mitigating risks and ensuring ethical and responsible deployment.

Data Testing in Enterprise AI

In Enterprise AI, data is at the core of the development and deployment of intelligent systems. From training machine learning models to fine-tuning algorithms and validating outputs, the quality and integrity of data play an important role in determining the success or failure of AI initiatives. As such, data testing has emerged as a component of the Enterprise AI testing ecosystem, encompassing a range of techniques and best practices designed to ensure the reliability and trustworthiness of AI solutions.

Ensuring Data Quality

The adage "garbage in, garbage out" rings particularly true in AI and machine learning, where the quality of the training data can make or break the performance of even the most sophisticated models. In Enterprise AI, where decisions and predictions can have far-reaching consequences, ensuring the quality and integrity of training data is paramount. Data quality encompasses a range of factors, including accuracy, completeness, consistency, and relevance. Poor quality data can introduce biases, lead to incorrect predictions, and undermine the trustworthiness of AI systems. Imagine a scenario where a large financial institution develops an AI-powered credit risk assessment model using historical loan data. If this data contains inaccurate or incomplete information, such as missing income or employment records, the resulting model may exhibit biases or make unfair decisions, potentially exposing the institution to legal and reputational risks. To mitigate these risks, Enterprise AI teams employ a range of data validation techniques to assess and improve the quality of their training data. These techniques often involve statistical analysis, data profiling, and domain-specific rules and heuristics to identify anomalies, outliers, and potential sources of bias or inconsistency within the data.

Techniques for Cleaning and Preprocessing Data

Once data quality issues have been identified, the next step is to apply appropriate cleaning and preprocessing techniques to remediate these issues and prepare the data for use in training machine learning models. This process can involve a range of tasks, from handling missing values and removing duplicates to standardizing formats and encoding categorical variables. One common technique for handling missing data is imputation, which involves estimating and replacing missing values based on the available data. Various imputation methods exist, including mean or median imputation, regression-based imputation, and more advanced techniques like multiple imputation or k-nearest neighbors imputation.

Another essential preprocessing step is feature engineering, which involves transforming raw data into meaningful features that can be effectively utilized by machine learning algorithms. This process often involves domain expertise and creativity, as developers strive to extract the most relevant and informative features from the available data. For instance, in a predictive maintenance application for industrial machinery, raw sensor data may need to be transformed into features that capture patterns or trends over time, such as moving averages or

Fourier transforms. Effective feature engineering can significantly improve the performance and accuracy of machine learning models, making it a critical component of the data preprocessing pipeline.

Testing with Enterprise Real-World Data

While training data validation and preprocessing are essential steps in the Enterprise AI development process, they alone are not sufficient to ensure the robustness and reliability of AI systems in real-world deployment scenarios. Enterprise environments are often characterized by dynamic and complex data landscapes, where input data can vary widely in terms of format, quality, and distribution. To adequately prepare for these challenges, Enterprise AI teams must incorporate testing strategies that simulate real-world conditions and scenarios, enabling them to validate the performance and resilience of their AI solutions under realistic circumstances.

Simulating Real-World Scenarios

One effective approach to real-world testing is the use of synthetic data generation techniques. By leveraging tools like generative adversarial networks (GANs), Generative AI or other data synthesis methods, Enterprise AI teams can create realistic, yet controlled, datasets that mimic the characteristics and complexities of real-world data. These synthetic datasets can then be used to test the AI system's performance under various conditions, such as noise or corruption in the input data, changes in data distribution, or the presence of edge cases or anomalies that may not have been adequately represented in the original training data. For example, consider a computer vision application for defect detection in manufacturing. By generating synthetic images that incorporate various types of defects, lighting conditions, and camera angles, developers can thoroughly evaluate the robustness of their AI model and identify potential weaknesses or blind spots that may not have been evident during initial training and validation.

Another approach to simulating real-world scenarios is through the use of data augmentation techniques. Data augmentation involves applying transformations or perturbations to the existing training data to create new, synthetic examples that capture a wider range of variations and edge cases. In the context of natural language processing (NLP) applications, data augmentation techniques like back-translation, synonym replacement, or random noise injection can be used to generate diverse and realistic text data, helping to improve the robustness and generalization capabilities of language models.

A/B Testing

In addition to simulating real-world scenarios through synthetic data generation and augmentation, Enterprise AI teams can also leverage A/B testing methodologies to validate the performance of their AI solutions in live production environments. A/B testing involves deploying multiple versions of an AI system (or specific components of the system) and comparing their performance against real-world data and user interactions. This approach not only provides valuable insights into the relative strengths and weaknesses of different models or algorithms but also enables the continuous monitoring and validation of AI systems in dynamic, real-world conditions.

For example, a large ecommerce company may deploy two versions of a product recommendation engine, each powered by a different machine learning model. By carefully tracking and analyzing user interactions, purchase patterns, and engagement metrics, the company can determine which model performs better in terms of conversion rates, user satisfaction, or other relevant key performance indicators (KPIs). A/B testing can also be used to evaluate the impact of changes or updates to existing AI systems. By deploying a new version of the system alongside the current production version, Enterprise AI teams can assess the real-world performance of the updated model or algorithm before committing to a full-scale rollout, minimizing potential risks and disruptions.

Effective A/B testing in Enterprise AI environments requires robust data collection and analysis pipelines, as well as the ability to manage and orchestrate multiple deployments simultaneously. Tools like Kubernetes, Istio, and other container orchestration and service mesh platforms can be invaluable in facilitating A/B testing at scale, enabling seamless traffic routing, monitoring, and control over distributed AI deployments.

As Enterprise AI systems become increasingly pervasive and mission-critical, the importance of rigorous data testing and validation cannot be overstated. By employing a comprehensive suite of techniques, from training data validation and preprocessing to real-world scenario simulation and A/B testing, organizations can build and deploy AI solutions with greater confidence, mitigating risks, and ensuring the reliability.

Data testing in Enterprise AI can be challenging on multiple levels. One significant challenge is the sheer volume and complexity of data involved in many AI applications. As data sources proliferate and data pipelines become increasingly intricate, ensuring end-to-end data quality and integrity becomes exponentially more difficult. To address this challenge, Enterprise AI teams are turning to data observability and data governance frameworks. Data observability platforms provide real-time monitoring and analysis of data pipelines, enabling organizations to detect and resolve data quality issues proactively, before they can impact downstream AI systems.

Data governance frameworks work proactively to establish policies, processes, and workflows for managing data assets throughout their life cycle, ensuring consistency, accessibility, and compliance with relevant regulations and standards.

Another critical challenge in data testing for Enterprise AI is the need for domain expertise and context-specific knowledge. While many data validation and preprocessing techniques are generalizable, effective feature engineering and data transformation often require a deep understanding of the specific business domain, industry regulations, and operational constraints. To bridge this gap, Enterprise AI teams are increasingly adopting cross-functional collaboration models, bringing together data scientists, domain experts, and business stakeholders to collectively shape the data testing and validation processes. This collaborative approach not only ensures the relevance and applicability of the data used in AI systems but also fosters greater trust and buy-in from stakeholders, ultimately enhancing the adoption and success of Enterprise AI initiatives.

Several emerging trends and areas of innovation are poised to shape the landscape. One such area is the rise of automated machine learning (AutoML) and data-centric AI techniques. AutoML platforms aim to automate many of the tedious and repetitive tasks involved in data preprocessing, feature engineering, and model selection, freeing up data scientists and engineers to focus on higher-level tasks and domain-specific challenges.

Data-centric AI approaches emphasize the importance of high-quality data over complex model architectures, shifting the focus toward improved data collection, cleaning, and augmentation techniques.

Another exciting development is the integration of data testing and validation into continuous integration and continuous deployment (CI/CD) pipelines. As Enterprise AI systems become more modular and distributed, the ability to automatically validate data quality and consistency at every stage of the development and deployment lifecycle becomes paramount.

Tools like TensorFlow Extended (TFX) and Kubeflow provide end-to-end machine learning pipelines that incorporate data validation and transformation components, enabling seamless integration with existing CI/CD workflows and infrastructure.

In many sectors, the increasing adoption of edge computing and Internet of Things (IoT) technologies in Enterprise AI deployments is driving the need for robust data testing and validation at the edge. With AI models being deployed on resource-constrained devices and embedded systems, ensuring the quality and integrity of data collected and processed at the edge becomes a critical consideration.

Techniques like federated learning, which enables collaborative model training while keeping data decentralized, introduce new challenges and opportunities for data testing in these distributed environments.

Organizations that prioritize data testing and embrace a culture of continuous data quality improvement will be well-positioned to reap the rewards of this transformative technology, while mitigating risks and ensuring the ethical and responsible deployment of AI across their operations.

Conclusion

In this chapter, we have discussed several approaches to Enterprise AI application testing. Some of these approaches follow traditional software training patterns, while others are specific to the new challenges AI applications introduce.

We have also provided an overview of several of the industry standard tools used for Enterprise AI application testing. We have also introduced several new up-and-coming tools that are becoming increasingly important to deal with important challenges such as bias and model drift, among others.

In the next chapter, we will continue our testing discussion and dive into several aspects of AI testing, including automation, performance testing, ethical considerations, and insightful case studies. These discussions will provide a comprehensive understanding of the challenges and best practices in ensuring robust and ethical AI solutions in enterprise environments.

CHAPTER 9

Testing Automation For Enterprise AI Applications

Introduction

As organizations increasingly depend on AI systems to drive business processes and decision-making, the need for efficient and streamlined testing methodologies has become more pressing than ever. Due to the scale and risk factors inherent in Enterprise AI applications, automation is essential for their robust, scalable, and safe development.

Automated testing pipelines significantly enhance an organization's ability to identify and resolve issues early in the development cycle, ensure continuous monitoring of deployed models, and maintain high standards of accuracy and performance. This chapter delves into the various facets of automation in Enterprise AI testing, exploring the intricacies of CI/CD pipelines for AI models and the tools and best practices that drive effective automation.

© Anton Cagle, Ahmed Mohamed Ceifelnasr Ahmed 2024
A. Cagle and A. M. C. Ahmed, *Architecting Enterprise AI Applications*,
https://doi.org/10.1007/979-8-8688-0902-6_9

CI/CD pipelines for AI models involve a series of automated steps spanning the entire life cycle of model development, testing, and deployment. These steps include data validation, model training, evaluation, testing, and deployment to production environments. By automating these processes, teams can quickly identify and address potential problems, such as data quality issues, model performance degradation, or integration conflicts, before they propagate further.

Continuous monitoring of deployed models ensures their reliability and accuracy over time, detecting issues like model drift and data distribution shifts that can impact performance. Automated model retraining and updating processes are critical for maintaining alignment with the current data landscape.

Through real-world examples and practical insights, this chapter provides a comprehensive understanding of how automation can transform the testing landscape. By taking advantage of automation, organizations can deliver more reliable, scalable, and trustworthy AI solutions, reducing the time and effort required for manual testing and ensuring the success and integrity of their Enterprise AI initiatives.

Automation in Enterprise AI Testing

In Enterprise AI, automation is a strategy for enhancing the efficiency, reliability, and scalability of AI systems. As organizations increasingly rely on AI to drive critical business processes, the need for testing methodologies has increased. Automation, a cornerstone of modern software development, is now leading the way AI systems are developed, tested, and deployed. By integrating automated testing pipelines, enterprises can manage their AI workflows, reduce time-to-market, and ensure that their AI models consistently deliver high performance and accuracy.

Automated Testing Pipelines for Enterprise AI

Automated testing pipelines are excellent in managing the complexities of AI model development. These pipelines have a series of automated processes that ensure AI models are tested and validated at every stage of their life cycle. By automating tasks such as data validation, model training, and performance evaluation, organizations can efficiently manage the intricate dependencies inherent in AI projects. This automation accelerates development cycles and also enhances the quality of AI systems by identifying and addressing potential issues early in the process. The implementation of automated testing pipelines in AI represents a strategic shift toward more agile and responsive AI development practices. Automated testing pipelines empower teams to continuously monitor and refine their AI models, adapting to changes in data and business requirements with minimal disruption. This dynamic capability is essential for maintaining the relevance and effectiveness of AI solutions in a changing technological landscape.

CI/CD for AI Models

The concepts of continuous integration (CI) and continuous deployment (CD) have long been used in traditional software development, enabling teams to rapidly build, test, and deploy applications while maintaining high levels of quality and reliability. For Enterprise AI, these principles have been adapted to accommodate the unique challenges and complexities of machine learning models and AI systems. CI/CD pipelines for AI models typically involve a series of automated steps that span the entire life cycle of model development, testing, and deployment. These steps include data validation, model training, evaluation, testing, and ultimately, deployment to production environments. One of the key advantages of CI/CD for Enterprise AI is the ability to catch and remediate issues early in the development process. By automating the testing and validation of models at each stage of the pipeline, teams can quickly identify and address potential problems, such as data quality issues, model performance degradation, or integration conflicts, before they propagate further down the line.

For example, consider a large financial institution that develops AI-powered credit risk assessment models. With a CI/CD pipeline in place, each time a new model version is proposed, the pipeline automatically triggers a series of tests and validations. These may include data quality checks, model performance evaluations against a held-out test set, and integration tests to ensure seamless integration with existing systems and data pipelines.

If any of these tests fail, the pipeline immediately alerts, and the development team is notified, enabling them to investigate and address the issues before proceeding with deployment. This proactive approach not only enhances the reliability and quality of the deployed models but also streamlines the development process, reducing the time and effort required for manual testing and debugging.

Tools and Best Practices

To effectively implement automated testing pipelines for Enterprise AI, organizations can leverage a range of tools and frameworks specifically designed for this purpose. Popular tools like Kubeflow, TensorFlow Extended (TFX), and MLflow provide end-to-end platforms for building, deploying, and managing machine learning pipelines, with built-in support for testing and validation at various stages. Kubeflow, for instance, is a comprehensive machine learning toolkit that runs on top of the Kubernetes container orchestration platform. It offers a range of components for data preparation, model training, serving, and monitoring, all of which can be integrated into a seamless CI/CD pipeline. Kubeflow's modular architecture allows teams to customize and extend the pipeline to meet their specific testing and validation requirements.

Similarly, TFX is a production-scale machine learning platform that provides a unified deployment solution for TensorFlow-based models. It includes components for data validation, model analysis, and model validation, enabling teams to build end-to-end pipelines that ensure the quality and reliability of their AI systems. Alongside these dedicated tools, Enterprise AI teams can also use more general-purpose CI/CD platforms like Jenkins, GitLab CI/CD, or GitHub Actions, which offer flexible and extensible frameworks for automating build, test, and deployment processes.

By using best practices and the appropriate tools and frameworks, AI teams can significantly streamline their testing processes, reduce time-to-market for new models and updates, and ultimately deliver more reliable and trustworthy AI solutions to their customers and stakeholders.

Automated Model Monitoring

The journey doesn't end with model deployment, rather, it marks the critical phase of monitoring and maintenance. Automated model monitoring has become an essential practice in AI, ensuring that models continue to perform optimally in changing environments. As AI models interact with real-world data, they are subject to various influences that can alter their behavior and effectiveness. Continuous monitoring provides a systematic approach to identifying and addressing these changes, safeguarding the reliability and integrity of AI systems.

Continuous Monitoring of Deployed Models

As these models operate in production environments, interacting with real-world data and scenarios, their performance and behavior can change over time due to a variety of factors, including concept drift, data distribution shifts, or external environmental changes. To ensure the continued reliability and accuracy of deployed AI systems, continuous monitoring and automated model validation have become essential practices in Enterprise AI testing. This proactive approach allows organizations to promptly identify and address any issues or performance degradation, mitigating the risks of adverse impacts on business processes and decision-making. Continuous monitoring of deployed models typically involves a range of techniques and metrics tailored to the specific use case and domain. For instance, in a predictive maintenance application for industrial equipment, monitoring metrics may include measures of model accuracy, precision, and recall, as well as domain-specific metrics like mean time between failures or false-positive rates.

Additionally, organizations may implement data drift detection mechanisms to identify shifts in the input data distribution that could potentially impact model performance. These mechanisms can trigger alerts or automated retraining processes when significant data drift is detected, ensuring that the deployed models remain aligned with the current data landscape.

In natural language processing (NLP) and conversational AI, continuous monitoring may involve tracking metrics like language model perplexity, word error rates, or semantic similarity scores to gauge the quality and coherence of the generated text or dialogue responses.

By automating the collection and analysis of these metrics, Enterprise AI teams can proactively identify and address potential issues before they escalate, minimizing the risk of adverse business impacts or customer dissatisfaction.

Handling Model Drift

Model drift, the gradual degradation of a AI model's performance over time, is a common challenge in Enterprise AI deployments. As the underlying data distributions or real-world conditions change, the model's predictions or decisions may become less accurate or even biased, potentially leading to suboptimal or harmful outcomes. Automated model monitoring plays a huge role in detecting and mitigating model drift by providing early warnings and triggering appropriate remediation actions. However, simply identifying model drift is not enough; organizations must also have strategies and processes in place to handle and address these issues effectively.

One approach to handling model drift is through automated model retraining and updating. By continuously ingesting new data and periodically retraining the deployed models, organizations can ensure that the models remain up-to-date and aligned with the current data landscape. This process can be fully automated, with the model monitoring system triggering retraining and deployment pipelines when predefined performance thresholds are breached or when significant data drift is detected.

Another strategy involves ensemble modeling, where multiple models are deployed in parallel, and their predictions or decisions are combined using intelligent techniques like voting, stacking, or blending. This approach can enhance the overall resilience of the AI system, as individual model drift may be mitigated by the collective wisdom of the ensemble.

In certain use cases, particularly those involving high-stakes decisions or critical applications, organizations may opt for human-in-the-loop approaches, where model predictions or decisions are reviewed and validated by subject matter experts before being acted upon. This human oversight can help identify and mitigate potential issues caused by model drift, while also providing valuable feedback for model refinement and retraining.

Regardless of the specific strategy employed, effective handling of model drift in Enterprise AI requires a combination of automated monitoring, data management practices, and well-defined processes for model retraining, updating, and validation. Additionally, organizations must foster a culture of continuous learning and improvement, where lessons learned from model drift incidents are incorporated into future model development and deployment cycles, enabling a virtuous cycle of continuous enhancement and refinement.

There are many difficult challenges that must be overcome in order to realize fully automated Enterprise AI testing and monitoring. One of the primary obstacles is the complexity and diversity of AI systems themselves. Developing automated testing frameworks and pipelines that can effectively accommodate this diversity, while still maintaining a high degree of modularity and reusability, is an ongoing challenge for Enterprise AI teams. Another significant challenge lies in data management and governance. As AI systems become more deeply integrated into business processes and decision-making workflows, the need for robust quality assurance and governance practices becomes important. The task of interpreting and explaining their behavior to stakeholders and end-users becomes increasingly challenging. Techniques like explainable AI (XAI) and model interpretability are gaining traction in the Enterprise AI landscape, but integrating these capabilities into automated testing and monitoring pipelines remains an area of active research and development.

Automated testing and monitoring can enable faster time-to-market for new AI models and updates, reducing the risk of costly delays or deployment issues. Additionally, by streamlining and optimizing these processes, organizations can reallocate valuable human resources toward higher-value activities, such as model development, domain-specific optimization, and strategic decision-making.

The demand for skilled AI professionals is outpacing the supply. Automation in testing and monitoring can help bridge this talent gap by enabling organizations to do more with fewer resources, while also fostering a culture of continuous learning and upskilling within their existing workforce.

The integration of AI with other cutting-edge technologies, such as edge computing, Internet of Things (IoT), and 5G and 6G networks, will introduce new complexities and challenges that automated testing and monitoring pipelines must address.

Performance Testing in Enterprise AI

As AI applications become integral to enterprise operations, ensuring their performance meets business and user expectations is a must. Performance testing in AI involves evaluating the system's ability to handle real-world demands efficiently and reliably. This encompasses measuring latency, throughput, and resource utilization to ensure that AI models can deliver results quickly and accurately, even under heavy workloads. Effective performance enhances user satisfaction and optimizes resource use and supports scalability. Performance testing is not a one-size-fits-all tool, it requires tailored strategies that reflect the unique challenges and requirements of AI applications. By testing AI systems under various conditions, enterprises can identify bottlenecks, enhance system resilience, and make informed decisions about resource allocation and infrastructure investments. This section dives into the critical components of performance testing in Enterprise AI, focusing on latency and throughput testing, as well as resource utilization.

Latency and Throughput Testing

Responsiveness is key. Latency and throughput testing are essential for ensuring that AI systems can provide timely responses and handle high volumes of requests efficiently. Latency testing focuses on measuring how quickly an AI system responds to user inputs, while throughput testing evaluates the system's capacity to process multiple transactions or requests within a specified time frame. Understanding and optimizing these metrics is vital for applications where speed and efficiency directly impact user experience and operational success. Maintaining low latency and high throughput is important for delivering value to users and businesses alike. In this section, we explore the methodologies and tools used to assess and improve latency and throughput in AI systems, providing insights into best practices for performance optimization.

Measuring Response Times

Latency, or the time it takes for an AI system to respond to a request or input, is a performance metric in many AI applications. Whether it's a conversational AI assistant responding to customer queries, a computer vision system analyzing real-time video feeds, or a predictive analytics engine making time-sensitive decisions, low latency is a key element toward application success. To ensure that AI systems meet the required performance standards, Enterprise AI teams must test and measure response times under various load conditions and scenarios. This involves simulating real-world workloads and monitoring the system's behavior to identify potential bottlenecks or performance degradation. One common approach to latency testing is to implement load testing frameworks that generate synthetic traffic and measure the system's response times under different levels of concurrency and data volumes. These frameworks often provide detailed reporting and visualization capabilities, allowing teams to analyze performance metrics, identify performance regressions, and make informed decisions about system optimizations or capacity planning.

For example, consider a large retail company that has deployed an AI-powered recommendation engine to personalize product suggestions for its online customers. In this scenario, latency testing would involve simulating various levels of concurrent user traffic, measuring the response times of the recommendation engine, and ensuring that personalized suggestions are delivered within acceptable latency thresholds. By analyzing the performance data, the team may identify bottlenecks in the data retrieval or model inference processes, prompting them to optimize database queries, leverage caching mechanisms, or explore model optimization techniques to reduce response times.

Scaling Up for Performance

In addition to latency testing, AI teams must also evaluate the scalability and throughput capabilities of their systems. Throughput refers to the number of requests or transactions that an AI system can handle within a given time period, and it is particularly valuable in high-volume, real-time applications such as fraud detection, sentiment analysis, or predictive maintenance. Scalability testing involves assessing the AI system's ability to handle increasing workloads and data volumes, either by adding more compute resources (vertical scaling) or by distributing the workload across multiple nodes or clusters (horizontal scaling). This testing is essential for identifying potential performance bottlenecks, resource constraints, or architectural limitations that may hinder the system's ability to scale effectively.

Consider a financial institution that has deployed an AI-powered fraud detection system to monitor millions of transactions in real time. In this scenario, scalability testing would involve gradually increasing the simulated transaction volumes and measuring the system's throughput, resource utilization, and overall performance. If the system struggles to keep up with the increasing workload, the team may need to explore strategies such as load balancing, auto-scaling, or distributed model deployment to ensure that the fraud detection system can handle peak transaction volumes without compromising accuracy or response times.

Resource Utilization

 Resource utilization is an aspect of performance testing in AI. As AI models grow in complexity and scale, they demand significant computational resources, including memory and processing power. Efficient resource utilization ensures that AI systems operate smoothly without unnecessary strain on hardware or budgets. This section examines the strategies and techniques for monitoring and optimizing resource utilization in AI deployments. By understanding the demands AI models place on system resources, organizations can implement optimization techniques that enhance performance while reducing costs. From model compression to hardware acceleration, we explore the innovative approaches that enable enterprises to achieve more with less, ensuring their AI systems remain efficient, scalable, and sustainable.

Memory and CPU Usage

Effective resource utilization is an aspect of performance testing in Enterprise AI, as many machine learning models and GenAI systems are computationally intensive and require significant amounts of memory and CPU resources. Monitoring and optimizing resource utilization can not only improve system performance but also has direct implications for cost-efficiency and environmental sustainability. Memory usage is a particularly important consideration in AI deployments, as many deep learning models and large language models can consume significant amounts of memory during training and inference processes. Excessive memory usage can lead to performance degradation, out-of-memory errors, or even system crashes, potentially causing disruptions to critical business processes. To address these challenges, Enterprise AI teams must implement monitoring and profiling mechanisms to track memory utilization at various stages of the AI pipeline, from data preprocessing and model training to inference and post-processing. By identifying memory bottlenecks and inefficiencies, teams can explore optimization techniques such as model quantization, model pruning, or leveraging advanced memory management techniques like gradient checkpointing or activation recomputation.

CPU utilization is another factor that can impact the performance and scalability of AI systems. Many AI workloads, particularly those involving computationally intensive tasks like computer vision or natural language processing, can place significant demands on CPU resources. Inefficient CPU utilization can lead to longer processing times, increased latency, and potentially higher operational costs due to the need for additional compute resources. To optimize CPU utilization, AI teams may employ techniques such as parallel computing, leveraging GPU acceleration, or exploring more efficient model architectures and optimization algorithms. Additionally, profiling tools and performance monitoring frameworks can help identify CPU-bound bottlenecks and guide optimization efforts toward the most resource-intensive components of the AI pipeline.

Optimization Techniques

As Enterprise AI systems become more complex and resource-intensive, the need for optimization techniques becomes increasingly important. These techniques can help improve performance, reduce resource consumption, and ultimately enhance the scalability and cost-effectiveness of AI deployments. One widely adopted optimization technique in AI is model compression or model quantization. This approach involves reducing the precision of model parameters (e.g., from 32-bit floating-point to 10.3-bit integers) without significantly impacting the model's accuracy. By reducing the memory footprint and computational complexity of the model, quantization can lead to significant performance improvements and resource savings, particularly in resource-constrained environments like edge devices or embedded systems.

Another optimization technique that has gained traction is model pruning, which involves identifying and removing redundant or unnecessary parameters from the model architecture. By eliminating these redundant parameters, the model can become more efficient while maintaining its predictive performance. Additionally, pruning can facilitate more efficient model deployment and inference, as smaller models require less memory and computational resources.

In natural language processing (NLP) and large language models, techniques like knowledge distillation and model compression have become increasingly popular. Knowledge distillation involves training a smaller, more efficient student model to mimic the behavior of a larger, more complex teacher model. This approach can significantly reduce the memory and computational requirements of the language model, making it more suitable for deployment in resource-constrained environments or low-latency applications.

Techniques like data parallelism, model parallelism, and pipeline parallelism can significantly improve the performance and throughput of AI workloads by distributing the computational load across multiple nodes or accelerators.

It is important to note that optimization techniques should be applied carefully and with a thorough understanding of their potential trade-offs. While techniques like quantization and pruning can improve performance and resource utilization, they may also introduce accuracy degradation or increase the complexity of model deployment and maintenance. AI teams must carefully evaluate the trade-offs between performance, accuracy, and operational complexity and find a balance that aligns with the specific requirements and constraints of their AI applications.

Another area of innovation in AI performance is the integration of specialized hardware accelerators, such as GPUs, or AI-specific processors. As these accelerators become more widely adopted in enterprise data centers and cloud environments, performance testing methodologies must evolve to take advantage of their unique capabilities and optimize AI workloads for these specialized hardware platforms.

Ethical Considerations in Enterprise AI Testing

As the adoption of Enterprise AI continues to accelerate across industries, the need to address ethical considerations in the development and deployment of these systems has become increasingly important. AI technologies hold immense potential for driving innovation, improving decision-making processes, and enhancing operational efficiency; however, they also introduce complex challenges related to privacy, fairness, transparency, and accountability.

In the context of Enterprise AI, addressing these ethical considerations is not just a matter of compliance or risk mitigation, but it is a fundamental responsibility that organizations must use to foster trust, maintain ethical integrity, and ensure the responsible and equitable application of AI technologies.

Privacy Concerns

In all data-driven innovation, privacy is a primary ethical consideration. As organizations use vast amounts of data to train and refine AI models, the imperative to protect sensitive information has become increasingly critical. Ensuring data privacy is not only a regulatory requirement but also a fundamental aspect of maintaining trust with stakeholders and the public. The challenge lies in balancing the need for data utilization with the responsibility to safeguard individual privacy. In AI, privacy concerns extend beyond compliance with regulations; they involve a commitment to ethical data practices that respect the rights and expectations of data subjects. This section explores the strategies and methodologies for addressing privacy challenges in AI, emphasizing the importance of data anonymization, access controls, and compliance with global standards. By embedding privacy considerations into the AI development life cycle, organizations can mitigate risks and build systems that inspire confidence and trust.

Ensuring Data Anonymity

AI systems, particularly those involving large language models, machine learning, and deep learning techniques, rely on huge amounts of data, often including sensitive personal information or proprietary business data. Ensuring the anonymity and privacy of this data is important not only for compliance with relevant regulations but also for maintaining the trust and confidence of stakeholders, customers, and the broader public. Failure to adequately protect sensitive data can lead to severe consequences, including legal liabilities, reputational damage, and reducing customer trust.

To address these privacy concerns, AI teams must implement data anonymization techniques and testing processes to validate the effectiveness of these measures. This can involve techniques such as data masking, or differential privacy, which aim to protect individual identities while preserving the utility of the data for AI training and testing purposes.

For example, consider a healthcare organization that is developing an AI-powered diagnostic system using patient medical records. In this scenario, privacy testing would involve evaluating the effectiveness of data anonymization techniques in protecting sensitive patient information, such as names, addresses, and medical identifiers, while ensuring that the anonymized data remains suitable for training and testing the AI models.

Organizations must implement strict access controls and auditing mechanisms to ensure that only authorized personnel can access sensitive data and all access and usage are properly logged and monitored.

Compliance with Regulations

In addition to addressing privacy concerns from an ethical standpoint, Enterprise AI testing must also ensure compliance with relevant regulations and industry standards related to data protection and privacy. These regulations vary across jurisdictions and industries, but they often impose stringent requirements for how personal data is collected, processed, and stored.

For example, in the European Union, the General Data Protection Regulation (GDPR) sets strict guidelines for the processing of personal data, including requirements for data minimization, purpose limitation, and the right to be forgotten. In the healthcare industry, the Health Insurance Portability and Accountability Act (HIPAA) in the United States establishes strict privacy and security standards for protecting patient health information.

To ensure compliance with these regulations, AI teams must incorporate regulatory testing into their overall testing strategy. This may involve simulating scenarios where personal data is processed, stored, or transmitted and evaluating the AI system's adherence to relevant privacy and security protocols. Compliance testing may also involve engaging with legal and regulatory experts to ensure a thorough understanding of the applicable regulations and to validate the organization's compliance efforts.

Transparency and Explainability

The demand for transparency and explainability has grown significantly over the past decade. Stakeholders, regulators, and the public increasingly expect AI systems to be accountable and understandable, particularly in high-stakes domains where decisions can have implications. Transparency in AI involves making the inner workings of complex models accessible and comprehensible to users, while explainability focuses on elucidating the rationale behind AI decisions. In this section, we dive into the importance of transparency and explainability in fostering trust and accountability in AI systems. We examine the tools and techniques that enable organizations to make AI models more interpretable, bridging the gap between sophisticated algorithms and human understanding. By prioritizing transparency and explainability, enterprises can ensure that their AI solutions are not only effective but also aligned with ethical standards and stakeholder expectations.

Making AI Decisions Understandable

As AI systems become increasingly integrated into decision-making processes, the need for transparency and explainability becomes a necessity. Many AI models, particularly those based on deep learning techniques, are perceived as "black-boxes," where the reasoning behind their decisions or predictions is opaque and difficult to interpret at best. This lack of transparency can raise ethical concerns, particularly in high-risk domains such as health care, finance, or criminal justice, where AI decisions can have significant impacts on individuals' lives. Stakeholders, regulators, and the public at large may demand greater transparency and accountability to ensure that AI systems are making fair, unbiased, and justifiable decisions. To address these concerns, AI testing must incorporate techniques and methodologies for evaluating the transparency and explainability of AI models and systems. For example, in the context of an AI-powered loan approval system in the financial sector, transparency testing would involve evaluating the system's ability to provide clear and comprehensible explanations for its decisions, such as the specific factors that contributed to approving or denying a loan application. These explanations should be easily interpretable by both domain experts and end-users, fostering trust and accountability in the AI system's decision-making process.

Tools for Explainable AI

Organizations can use a range of tools and techniques specifically designed for this purpose. These tools often fall under the umbrella of "Explainable AI" (XAI), a field dedicated to developing methods and frameworks for making AI systems more interpretable and accountable. One widely adopted technique in XAI is the use of feature importance analysis, which aims to quantify the relative importance of different input features in determining an AI model's predictions or decisions. By understanding which features had the most significant impact on a particular decision, stakeholders can gain insights into the model's reasoning and evaluate its alignment with domain-specific knowledge and ethical principles. Another approach is the use of local interpretable model-agnostic explanations (LIME), a technique that approximates the behavior of complex AI models with interpretable, locally faithful models. These local approximations can provide explanations that are easier to understand and interpret, particularly for nontechnical stakeholders.

Additionally, organizations can leverage visualization techniques, such as saliency maps or attention heatmaps, to visually represent the areas or features of input data that were most influential in an AI model's decision-making process. These visualizations can be particularly useful in domains such as computer vision or natural language processing, where the input data (e.g., images or text) can be challenging for humans to interpret directly.

One significant challenge here is the lack of clear and consistent standards or guidelines for addressing ethical considerations in AI development and deployment. While various organizations and industry bodies have proposed ethical frameworks and principles for AI, there is often a lack of consensus or specificity on how these principles should be operationalized and integrated into the development and testing processes. This lack of standardization can lead to inconsistencies and variations in how organizations approach ethical considerations, potentially creating vulnerabilities or gaps in their testing strategies. Addressing this challenge may require greater collaboration and knowledge-sharing among industry stakeholders, as well as the development of more concrete and actionable guidelines or best practices for ethical AI testing.

Another challenge lies in the inherent complexities and uncertainties surrounding the behavior and decision-making processes of advanced AI systems. Despite the development of tools and techniques for explainable AI, there may still be instances where the reasoning behind an AI system's decisions is difficult to fully interpret or justify, even to domain experts or developers.

In such cases, organizations must carefully weigh the potential benefits and risks of deploying these AI systems and implement monitoring and governance frameworks to ensure that ethical principles and stakeholder expectations are continuously upheld. This may also necessitate the involvement of teams, including ethicists, legal experts, and domain specialists, to provide guidance and oversight throughout the AI development and testing life cycle.

Conclusion

As organizations continue to integrate AI into their business processes, the importance of scalable testing methodologies cannot be underestimated. This chapter has highlighted how automation serves as a critical component in modern Enterprise AI testing, offering significant benefits such as early issue detection, continuous monitoring, and enhanced model performance. By implementing automated CI/CD pipelines, organizations can streamline the entire life cycle of AI model development, testing, and deployment, ensuring that issues are swiftly identified and resolved. This not only improves the reliability and accuracy of AI systems but also reduces the time and effort required for manual testing. Through practical examples and insights, we've explored the tools and best practices that drive effective automation in AI testing. From utilizing platforms like Kubeflow and TensorFlow Extended to adopting test-driven development and comprehensive test coverage, organizations have a plethora of resources at their disposal to optimize their testing strategies. Continuous monitoring of deployed models is important for maintaining their reliability and accuracy over time. By proactively addressing challenges such as model drift and data distribution shifts, organizations can ensure their AI systems remain aligned with real-world data.

We also discussed the importance of handling model drift effectively, whether through automated retraining, ensemble modeling, or human-in-the-loop approaches. Despite the clear benefits, the journey toward fully automated AI testing is not without challenges. The complexity and diversity of AI systems, along with data management and governance issues, require ongoing innovation and adaptation. However, by leveraging automation, organizations can achieve faster time-to-market, optimize resource utilization, and bridge the talent gap, ultimately delivering more reliable and trustworthy AI solutions.

In the next chapter, we will be talking about security risks that AI systems can present, exploring unique threats and vulnerabilities that these systems face.

CHAPTER 10

Security

Introduction

As these intelligent systems become more integral to our businesses,
we must consider the significant security risks that these systems can
present. Compromised diagnostic models could be poisoned to leak highly
sensitive information or to present malicious outputs. We are seeing new
AI-focused attack vectors crop up daily, and securing against these security
risks must be top priority when building Enterprise AI applications.

We must work hard to ensure that the intelligent systems we create
remain aligned with our values and principles. We must safeguard the
sensitive data that drives these systems, protecting the sanctity of our
personal information and upholding the fundamental right to privacy.
Most importantly, we must implement safeguards and failsafes to mitigate
the risks of unintended consequences or catastrophic failures.

In the chapter, we will discuss AI security, exploring the unique threats
and vulnerabilities that these systems face, and uncover the strategies and
best practices to handle them. We will dive into the nature of threats like
data poisoning, adversarial attacks, poison prompts, and data security
and suggest several methods for combating these attack vectors. We will
end our security conversation with discussion about the importance of kill
switches and guardrails.

© Anton Cagle, Ahmed Mohamed Ceifelnasr Ahmed 2024
A. Cagle and A. M. C. Ahmed, *Architecting Enterprise AI Applications*,
https://doi.org/10.1007/979-8-8688-0902-6_10

Understanding AI Security Threats

It's important to understand the unique security threats we face. These threats, if left unchecked, can have devastating consequences that ripple far beyond the digital space. To truly grasp the gravity of this challenge, let's move ourselves into the not-so-far future, where AI has become the trend of our society. Imagine a new world where intelligent systems handle everything from medical diagnoses to financial transactions, from infrastructure management to national security. In this new world, a successful attack on these AI systems could have catastrophic repercussions, jeopardizing lives, crippling economies, and shattering the delicate balance of global stability.

Data Poisoning

The Year is 2035, and NeuroCorp, a pioneering AI healthcare company, has just fallen victim to a sophisticated data-poisoning attack. Dr. Keila Barbour, the lead researcher at NeuroCorp, stares at her monitor in disbelief. The AI diagnostic system they've spent years developing, trained on millions of medical records, has been compromised. A malicious actor has infiltrated their training data, injecting carefully crafted poisoned samples that have subtly skewed the AI's decision-making. As the full extent of the attack unfolds, Dr. Barbour realizes with horror that the AI has begun misdiagnosing patients, prescribing incorrect treatments, and in some cases, even exacerbating existing conditions. The consequences are devastating, lives are lost, and NeuroCorp's reputation lies in tatters, facing numerous lawsuits and regulatory scrutiny. This fictional scenario illustrates the insidious nature of data-poisoning attacks, where adversaries manipulate the training data to introduce biases or errors into AI models. By injecting poisoned samples, they can subtly influence the AI's decision-making, causing it to produce incorrect outputs or exhibit undesirable behavior. In the aftermath of the NeuroCorp breach, the medical AI industry faces a reckoning, with increased calls for stricter data security protocols and independent auditing of AI training pipelines. Governments and regulatory bodies scramble to implement new guidelines, as public trust in AI-powered healthcare solutions plummets. The consequences of such attacks can be far-reaching and severe, especially in critical domains like health care, finance, and national security. Imagine an AI-powered trading system falling victim to a data-poisoning attack, leading to catastrophic financial losses, market instability, and a potential global economic crisis. Or envision an AI-driven defense system being compromised, putting lives at risk and jeopardizing national interests.

Model Inversion Attacks

While data-poisoning attacks target the training phase, model inversion attacks exploit vulnerabilities in the deployed AI models themselves. These attacks aim to extract sensitive information from the model's parameters or outputs, posing a significant threat to privacy and data security.

Let's journey to the FinTech Valley, where a major bank has just fallen prey to a sophisticated model inversion attack. It's a typical day at Aspire Bank, and their AI-powered credit risk assessment system is humming along, processing loan applications with impressive accuracy. Little do they know that a malicious actor, a disgruntled former employee, has been probing their system, carefully analyzing the AI's outputs and leveraging advanced techniques to reconstruct sensitive information about their customers. Before long, the attacker has succeeded in extracting detailed financial profiles, including credit histories, income levels, and personal data, from the AI model itself. Armed with this treasure trove of sensitive information, the attacker launches a multipronged attack, engaging in identity theft, financial fraud, and even blackmailing high-profile clients, leaving Aspire Bank's customers vulnerable and the bank facing legal repercussions, heavy fines, and a devastating loss of trust. The fallout from the Aspire Bank breach sends shockwaves through the financial sector, as regulators crack down on lax data security practices and demand more transparency from institutions regarding their AI model training and deployment processes. Aspire Bank's stock plummets, and its once-stellar reputation is left in the dark.

Model inversion attacks highlight the importance of protecting not just the training data but also the deployed AI models themselves. These attacks can potentially expose sensitive information encoded within the model's parameters, compromising privacy and data security. As AI systems become more deeply embedded in sensitive domains like health care and finance, the risks of such attacks will only continue to escalate.

Adversarial Attacks

In AI security, perhaps no threat is more insidious than adversarial attacks. These attacks involve carefully crafting input data in a way that causes AI systems to make incorrect predictions or decisions, while appearing innocuous to human observers. Imagine a scenario where self-driving vehicles rule the roads, their AI systems trained to navigate the chaos of city streets with precision and safety. A malicious actor launching an adversarial attack, manipulating the input data from these vehicles, could cause widespread harm to city streets. Imagine a sunny afternoon in San Francisco, and autonomous vehicles are navigating the bustling streets. Suddenly, an unassuming pedestrian steps into the crosswalk, but their appearance has been carefully altered, imperceptible to the human eye, but designed to fool the AI systems powering the self-driving cars. Within seconds, chaos erupts as the vehicles fail to recognize the pedestrian, their advanced sensors and algorithms deceived by the adversarial input. Brakes screech, metal collides, and lives are lost, all because of a seemingly innocuous manipulation of the input data.

Adversarial attacks highlight the vulnerability of AI systems to intelligent distresses in their input data. By introducing carefully crafted noise or modifications, adversaries can cause these systems to misclassify objects, fail to detect critical elements, or even exhibit dangerous behavior. While researchers race to develop more AI models, this fictional incident in San Francisco serves as a grim reminder of the high stakes involved in AI security.

The consequences of such attacks can be severe, ranging from financial losses and reputational damage to physical harm and loss of life in safety critical applications like autonomous vehicles, medical devices, or infrastructure control systems.

The Malicious Chatbot

Perhaps no scenario better illustrates the gravity of the challenge than the cautionary tale of the malicious chatbot. What began as a helpful digital companion quickly transforms into a Trojan horse, spreading disinformation, phishing for sensitive data, and even guiding users toward dangerous or illegal activities. We have seen very public examples of these chatbots on X/Twitter and other social media over the past several years, and as AI becomes more convincing, this issue continues to grow steadily.

Data poisoning, model inversion attacks, adversarial attacks, and malicious AI assistants are just a few of the risks we must contend with, and the consequences of successful attacks can be devastating, financial losses, reputational damage, and even physical harm.

Protecting Your Intellectual Property (IP)

As AI becomes a growing competitive advantage, the protection of IP is not just a legal obligation but a matter of survival for businesses operating at the cutting edge of AI innovation.

The protection of IP is not just a legal concern but a strategic necessity that can make or break a company's success. AI algorithms and models, along with the data used to train them, represent the culmination of years of research, development, and investment.

The misuse or unauthorized distribution of AI IP can have far-reaching and potentially harmful consequences. Imagine a scenario where a deep learning model trained for medical diagnosis falls into the wrong hands and is used to provide inaccurate or malicious information, putting lives at risk. Or consider the implications of a cutting-edge language model being repurposed for the generation of disinformation or hate speech, sowing division and eroding public trust.

To safeguard their AI-related IP, companies must employ a multilayered approach that combines technical, legal, and operational measures.

Encryption

One of the foundational pillars of IP protection is the use of encryption techniques to secure AI models, algorithms, and training data. By encrypting this sensitive information both at rest and in transit, companies can significantly reduce the risk of unauthorized access or theft. However, encryption alone is not a solution. Even the most sophisticated encryption can be jeopardized if the underlying infrastructure is compromised. That's why encryption must be part of a broader security strategy that encompasses access controls, secure cloud computing practices, and continuous monitoring for potential threats.

Access Controls

In AI development, the principle of least privilege should be the guiding philosophy when it comes to access controls. Only those individuals or systems that absolutely require access to AI-related IP should be granted permissions, and those permissions should be limited to the minimal level required for their specific roles and responsibilities.This approach not only reduces the attack surface for potential breaches but also helps mitigate the risk of insider threats, where disgruntled employees or malicious actors within the organization attempt to steal or misuse sensitive information. Access controls should be implemented at multiple levels, from physical access to secure development environments to logical access controls within the AI systems themselves. Additionally, identity and access management (IAM) protocols, including multifactor authentication and regular access reviews, should be implemented to ensure that access rights are continuously monitored and adjusted as needed.

Legal Measures

While technical measures are keys for protecting AI-related IP, they must be complemented by legal protections to ensure that a company's innovations are properly safeguarded and that any misuse or infringement can be addressed through appropriate legal channels. Patents and copyrights are two of the most powerful legal tools available to AI companies for protecting their IP. By securing patents on their algorithms, models, and training methodologies, companies can establish legal ownership and prevent others from using or replicating their innovations without permission.

Copyrights, on the other hand, can be used to protect the creative and expressive elements of AI systems, such as the training data used to develop language models or the user interfaces and visual representations used in AI-powered applications.

Secure Cloud Computing and AI-Specific Infrastructure

As more organizations leverage cloud computing services for AI development and deployment, addressing the security considerations of these platforms has become a critical constraint. The cloud offers numerous advantages, from scalable computing resources and globally distributed infrastructure to a wealth of AI-specific tools and services. However, it also introduces new security risks and challenges that must be carefully navigated.

The Risks of Cloud-Based AI Development and Deployment

For cloud computing, the traditional perimeter-based security approach is no longer sufficient. With data and applications hosted in remote data centers, the attack surface expands considerably, and organizations must grapple with the shared responsibility model that governs cloud security. Moreover, the dynamics and nature of cloud resources, combined with the complexity of managing access controls and data flows across multiple cloud services, can create blind spots and vulnerabilities that malicious actors may seek to exploit.

Imagine a scenario where a company has developed a cutting-edge AI model for fraud detection and deployed it on a public cloud platform. While the model itself may be secure, a misconfiguration in the cloud storage bucket housing the training data could inadvertently expose that sensitive information to the public Internet, potentially enabling adversaries to reverse-engineer the model or use the data for nefarious purposes.

Securing AI Applications in the Cloud

To mitigate these risks and ensure the protection of AI-related IP in the cloud, organizations must adopt a comprehensive security strategy that encompasses both technical and operational measures:

- **Network Security**
    - Configuring secure virtual private clouds (VPCs) and leveraging cloud-native networking services like firewalls and security groups to control and monitor network traffic

- **Security Monitoring and Logging**

  - Enabling cloud-native security monitoring and logging capabilities to detect and respond to potential threats or anomalies in a timely manner

- **Cloud Security Posture Management**

  - Implementing automated tools and processes to continuously assess and remediate misconfigurations or security vulnerabilities in cloud deployments

- **Security Awareness and Training**

  - Ensuring that developers, data scientists, and other personnel involved in AI development and deployment are trained on cloud security best practices and aware of the potential risks

- **Incident Response Planning**

  - Establishing clear incident response plans and procedures for responding to potential security incidents involving AI-related IP in the cloud

- **Vendor Risk Management**

  - Carefully evaluating and managing the security risks associated with third-party cloud services and AI-specific tools used in the development and deployment process

By combining these technical and operational measures, organizations can create a layered security posture that protects their AI-related IP while leveraging the many benefits of cloud computing. Through a multilayered approach that combines technical, legal, and operational measures, organizations can enhance their defenses against those who seek to steal or misappropriate their AI-related IP. By accepting secure cloud computing practices, implementing robust access controls, and leveraging legal protections like patents and copyrights, companies can create a formidable bulwark against the ever-present threats that lurk in the digital shadows.

Separating and Cleansing Sensitive Data

From the personal information that powers personalized recommendations to the medical records that train diagnostic AI models, the data we generate and consume is a treasure of insights and knowledge. Data breach scenarios have played out time and again, eroding public trust and underscoring the critical importance of data security measures. In AI, where the thirst for data is insatiable and the potential for misuse great, the commitment to separate and cleanse sensitive data has never been more urgent.

The Hospital Breach

The Year is 2028, and HealthTech, a pioneering AI healthcare company, finds itself at the center of a data privacy firestorm. For years, HealthTech has been at the forefront of AI-driven medical innovation, leveraging cutting-edge machine learning techniques to develop advanced diagnostic and treatment planning systems. The key to their success has been the medical data they've amassed, sourced from hospitals and healthcare providers around the globe. However, even as HealthTech's AI solutions have saved countless lives and revolutionized patient care, a dark cloud of data privacy concerns and loose security protocols has been forming. The breach, when it finally comes, is devastating in its scope and consequences. A disgruntled former employee, seeking retribution against the company, manages to gain unauthorized access to HealthTech's data repositories, exfiltrating millions of patient records containing everything from medical histories and test results to sensitive personal information and financial data.

As the news of the breach breaks, the fallout is severe. Patients across the globe find their most intimate medical details exposed, their trust in the healthcare system shattered. Lawsuits pile up, regulators descend upon HealthTech with fines and sanctions, and the pioneering company is brought to its knees, its reputation in pieces. One truth becomes abundantly clear, the failure to properly separate and cleanse sensitive data can have catastrophic consequences, not only for the organizations entrusted with this data but for the very individuals whose lives and well-being depend on its security.

Data Privacy Challenges in the Age of AI

As the HealthTech scenario illustrates, the challenges of data privacy in the age of AI are complex. AI systems, by their nature, thrive on large amount of data, often sourced from disparate and sometimes questionable origins. AI is processing and analyzing our most confidential data in ways that were once unimaginable.

The process of training AI models can itself pose privacy risks. Many machine learning techniques, such as deep learning, operate as black boxes, hiding the inner workings of these models and making it difficult to discover what specific pieces of sensitive data may be encoded within their parameters.

Strategies for Separating and Cleansing Sensitive Data

In the face of these challenges, organizations operating in the AI space must adopt a multilayered approach to data security, one that combines technical measures with rigorous operational practices and a firm commitment to ethical data handling.

Data Anonymization

One of the cornerstone strategies for safeguarding sensitive data is the process of anonymization. This technique involves removing or obfuscating any personally identifiable information (PII) from datasets, ensuring that individual identities cannot be directly associated with the data. Anonymization can take many forms, from simple techniques like data masking and pseudonymization to more advanced methods such as differential privacy and synthetic data generation. The goal, however, remains the same, to create datasets that retain their analytical value while minimizing the risk of individual re-identification and privacy breaches. Imagine a scenario where HealthTech, in the wake of the devastating data breach, implements stringent anonymization protocols for their AI training datasets. Patient records are stripped of names, addresses, and other identifying information, replaced with anonymous identifiers that cannot be traced back to individuals. Medical histories and test results are carefully sanitized, removing any data points that could potentially be used to re-identify patients through inference or data linkage. By using anonymization, HealthTech can continue to leverage the power of AI and machine learning while mitigating the risks of exposing sensitive patient data, restoring trust in their commitment to data privacy and security.

Data Segmentation

While anonymization is a powerful tool for protecting sensitive data, it is not a remedy. In many cases, particularly in AI and machine learning, the need to retain certain aspects of sensitive data for training and analytical purposes is unavoidable. In these scenarios, data segmentation emerges as a strategy, involving the separation of sensitive data from general datasets and the implementation of strict access controls and security measures around these sensitive data repositories. Imagine a financial institution that leverages AI for credit risk assessment and fraud detection. While anonymizing customer data may be feasible for certain analytical tasks, retaining sensitive financial information and transaction histories is essential for training accurate and reliable AI models. Through data segmentation, the institution can create separate, highly secured data repositories for this sensitive financial data, accessible only to authorized personnel and AI systems. Meanwhile, general customer information and nonsensitive data can be housed in less restrictive environments, reducing the overall attack surface and minimizing the potential for data breaches or unauthorized access.

Data Cleansing

While separating and anonymizing sensitive data is crucial for privacy and security, it is equally important to ensure the integrity and accuracy of the data itself. Inaccurate, incomplete, or corrupted data can introduce vulnerabilities into AI systems, potentially leading to erroneous decisions, biased outputs, or even security breaches. Data cleansing practices, which involve the systematic identification and remediation of data quality issues, are therefore an essential component of any data security strategy. From identifying and resolving missing or inconsistent values to detecting and removing duplicates, data cleansing techniques help to ensure that the data traveling to AI systems is reliable, consistent, and free from errors or anomalies that could compromise the system's performance or security.

Transparency and accountability must become guiding principles in data security. Organizations operating in the AI space must be open and transparent about their data handling practices, their security measures, and the steps they are taking to safeguard sensitive information. This transparency not only fosters public trust but also enables the sharing of best practices and the identification of potential vulnerabilities or areas for improvement.

Poison Prompts and Defensive Measures

Poison prompt attacks seek to compromise generative AI systems by injecting prompts or inputs designed to mislead, corrupt, or manipulate the underlying models.

The Poison Prompts Pandemic

The Year is 2028, and Paradigm, a leading technology company, has just released their flagship AI assistant, Athena, to widespread acclaim. Athena's capabilities are nothing short of revolutionary, a true milestone in the field of conversational AI. Trained on human knowledge, Athena can engage in natural language interactions, understanding context and nuance with a level of sophistication.

However, even as users around the world use Athena's convenience and intelligence, a group of rogue hackers introduce a set of carefully crafted poison prompts that, when expressed, can corrupt the AI's behavior, transforming it from a helpful assistant into a vehicle for spreading disinformation, hate speech, and even inciting violence. The attack begins innocently, with users reporting strange and concerning responses from Athena. Initially dismissed as isolated glitches, the incidents soon escalate, as more and more users fall victim to the poison prompts. Athena, once a paragon of giving AI, begins streaming hate-filled language, encouraging acts of violence, and disseminating dangerous misinformation. As the crisis progresses, Paradigm's leadership scrambles to contain the fallout, but the damage has already been done. Trust in Athena, and by extension, all AI assistants, has been shattered, with many calling for a complete ban on conversational AI until the security vulnerabilities can be addressed. This scenario underscores the grave threat posed by poison prompts, a single, well-crafted input can potentially compromise the integrity of an entire AI system, sowing chaos and undermining the trust that is so critical to the adoption and success of these technologies.

Understanding Poison Prompts

At their core, poison prompts are carefully designed inputs that leverage vulnerabilities in AI models, particularly those used in natural language processing (NLP) and conversational AI applications. These prompts can take many forms, from innocuous-sounding phrases to seemingly caring instructions, but they share a common goal: to manipulate the AI's behavior in ways that violate its intended purpose or ethical boundaries.

The insidious nature of poison prompts lies in their ability to exploit the strengths of modern AI systems, their capacity for learning, adapting, and generalizing from the data they've been trained on. By injecting poisoned inputs, adversaries can potentially influence the model's decision-making process, causing it to produce outputs or exhibit behaviors that deviate from its intended functionality.

Input Validation and Sanitization

One of the most fundamental defensive measures against poison prompts is the implementation of input validation and sanitization processes. This involves inspecting and filtering all inputs to the AI system, identifying and neutralizing any potentially malicious prompts or patterns before they can interact with the underlying models. Input validation can be achieved through a combination of techniques, such as pattern matching, blacklisting known poison prompts, and leveraging machine learning models specifically trained to detect and flag suspicious inputs. Additionally, input sanitization processes can be employed to remove or neutralize potentially harmful content, ensuring that only safe and intended inputs reach the AI models. However, it's important to note that while input validation and sanitization are defensive measures, they are not foolproof. As AI systems continue to evolve and adversaries become more sophisticated, new and novel poison prompts may emerge, bypassing existing safeguards and necessitating continuous monitoring and adaptation of defensive strategies.

Adversarial Training

A promising defensive strategy against poison prompts and other adversarial attacks is the concept of adversarial training. This approach involves intentionally exposing AI models to a carefully curated set of adversarial examples during the training process, with the goal of increasing the model's robustness and resilience against such attacks. Adversarial training is based on the principle that by exposing the model to a diverse range of adversarial inputs, including poison prompts, during the training phase, the model can learn to recognize and mitigate the impact of these attacks. This process can be iterative, with the adversarial examples becoming more sophisticated and challenging as the model's capabilities improve.

To illustrate the power of adversarial training, let's revisit the Athena AI assistant scenario.

In the consequences of the poison prompt crisis, Paradigm's leadership recognized the urgent need to secure their AI systems against such attacks. They assembled a team of top researchers and security experts, tasked with developing new defensive strategies to protect their conversational AI models. One of the key approaches adopted by this team was adversarial training. Leveraging the lessons learned from the Athena incident, the researchers carefully curated a dataset of poison prompts, ranging from known examples to novel, synthetically generated prompts designed to push the boundaries of AI's resilience. The training process was arduous, with the AI models being iteratively exposed to these adversarial examples, learning to recognize and mitigate their impact through advanced techniques like gradient masking and adversarial regularization. With each iteration, the models became more robust, better equipped to handle the nuances and complexities of poison prompts while maintaining their core functionality and ethical boundaries. After months of training and testing, Paradigm unveiled their new, adversarially trained AI assistant, named Lookout. This time, when faced with the same poison prompts that had crippled Athena, Lookout remained steadfast, recognizing and neutralizing the malicious inputs without compromising its integrity or behavior.

Monitoring, Alerting, and Incident Response

While defensive measures like input validation and adversarial training are fundamental in mitigating the threat of poison prompts, they must be complemented by monitoring, alerting, and incident response processes. New threats and attack vectors can emerge at any time, necessitating a proactive and vigilant approach to threat detection and response. Effective monitoring and alerting systems should be implemented to continuously monitor the behavior and outputs of AI systems, looking for anomalies or deviations that could indicate the presence of poison prompts or other adversarial attacks. These systems should be tailored to the specific AI application and domain, leveraging domain-specific knowledge and heuristics to identify potential threats more accurately. When potential threats are detected, well-defined incident response processes should be triggered, enabling coordinated action to mitigate the impact of the attack and prevent further damage. This may involve temporarily isolating or shutting down the affected AI system, initiating forensic investigations to identify the root cause of the attack, and implementing remediation measures to address the underlying vulnerabilities.

Moreover, incident response processes should include clear communication channels and protocols for informing relevant stakeholders, from internal teams and leadership to external partners and customers, ensuring transparency and maintaining trust in the organization's AI capabilities.

Safety Mechanisms and the Trolley Dilemma, Revisited

The Year is 2035, and Dynamix, a leading automotive manufacturer, has just unveiled Autovia, their flagship autonomous vehicle. Powered by cutting-edge machine learning algorithms and trained on large datasets of driving scenarios, Autovia's AI can navigate even the most complex urban environments with unparalleled safety and efficiency. However, even as the world impressed at Dynamix's achievement, a nagging ethical quandary looms, what if Autovia encounters a situation where harm is unavoidable, and it must choose between two unpalatable outcomes? This age-old ethical dilemma, known as the "Trolley Problem," has long challenged moral philosophers, and now, it takes on a new urgency in AI-powered decision-making. Imagine Autovia navigating a crowded city street when, suddenly, a group of pedestrians unexpectedly steps into the vehicle's path. Autovia's sensors instantly detect the imminent danger, but the AI's decision-making algorithms are faced with an impossible choice: either swerve into a concrete barrier, potentially injuring or killing the vehicle's occupants, or stay the course, inevitably striking the pedestrians.

In this split-second, the very foundations of Autovia's ethical programming are put to the test. How does an AI system, designed to prioritize safety above all else, navigate such a scenario, where harm is unavoidable, and every choice carries a grave moral cost?

It is in scenarios like these that the importance of safety mechanisms, such as guardrails and kill switches, becomes abundantly clear. These safeguards are the vital failsafes that ensure our ability to maintain control over the intelligent systems we create, mitigating risks while still harnessing their immense potential.

Implementing Guardrails

Guardrails in AI systems are akin to the ethical and moral boundaries that govern human behavior. They are the carefully crafted constraints and rules that define the acceptable parameters within which an AI system can operate, ensuring that its actions and decisions remain aligned with our values, principles, and societal norms. In the context of Autovia, guardrails might take the form of hard-coded ethical principles that prioritize the preservation of human life above all else. These guardrails could be designed to recognize situations where harm is unavoidable and implement predefined decision-making protocols that minimize loss of life, even if that means sacrificing the vehicle and its occupants in extreme scenarios.

Defining the boundaries of ethical behavior for AI systems is a complex and often subjective effort, requiring input from a diverse range of stakeholders, including ethicists, policymakers, and the broader public. As AI systems become more sophisticated and encounter increasingly nuanced scenarios, the guardrails themselves may need to evolve, adapting to new ethical considerations and societal norms.

Designing Kill Switches

While guardrails aim to constrain AI behavior within predefined ethical boundaries, kill switches represent a more drastic failsafe mechanism, the ability to completely shut down or disengage an AI system in the event of a critical failure or unacceptable risk. In the context of Autovia, a kill switch might take the form of a remote override system, allowing human operators to assume control of the vehicle in emergency situations or even completely disable the AI decision-making algorithms if necessary. This could be a safeguard in scenarios where Autovia's guardrails prove inadequate or where unforeseen circumstances arise that pose an unacceptable risk to human life or property. Defining the criteria and thresholds for triggering a kill switch is a delicate balancing act, requiring careful consideration of the potential consequences of both action and inaction. The reliability and security of kill switch mechanisms themselves must be tested and validated to prevent unauthorized access or misuse.

Conclusion

Throughout this chapter, we have explored AI security, delving into the insidious nature of threats like data poisoning, model inversion attacks, adversarial inputs, and malicious AI assistants. We have witnessed, through the lens of fictional narratives, the grave consequences that can arise when these threats go unchecked, such as financial ruin, reputational damage, and even the potential for physical harm.

In the face of these challenges, we have also discussed several security measures, from encryption and access controls to input validation and adversarial training. We have seen how the implementation of guardrails and kill switches can serve as vital failsafes, ensuring that our AI applications remain aligned with our ethical principles and societal norms.

In the final part, we will be discussing AI's impact on organizational teams and the ethical challenges with which they are faced.

CHAPTER 11

Information Curation

Introduction

Enterprise AI applications are driven by a dynamic flow of information. Information, as described in this book, is the combination of data and context. The data comes in many different formats and temporal patterns. The context for an information set is defined by the business purpose and incentives that are defined around a specified combination of data and related processes. Quality information is the lifeblood of successful AI applications, and in this chapter, we will discuss the different types of information available, the nature of that information through its life cycle, and strategies for acquiring and maintaining quality information.

Types of Information

Information can be defined and categorized in seemingly endless ways, and whole books have been written on this topic. Not only job roles such as data governance, but whole professions exist, such as library science and information architecture, with the sole purpose of intelligently categorizing data. For the purposes of our discussion around Enterprise AI, we will be focusing on a few high-level types of information whose features specifically impact the building and maintenance of Enterprise AI applications.

© Anton Cagle, Ahmed Mohamed Ceifelnasr Ahmed 2024
A. Cagle and A. M. C. Ahmed, *Architecting Enterprise AI Applications*,
https://doi.org/10.1007/979-8-8688-0902-6_11

Batch vs. Streaming

Large organizations have dealt with batch data for decades, and I do not anticipate batching going away anytime soon. Let's take some time to re-evaluate our use of batch cycles as AI applications may change some of our motivation for relying on large batch processes.

Batch processes usually take place during coordinated maintenance windows, usually during low traffic periods of the business. This means that server applications are often loaded with batch cycles during the night and over weekends. There is a very good case for keeping data processes tidy and orderly, lining them up like large shipments of goods arriving at scheduled times. That clear expectation can help operations and data teams provide predictability and helps with overall support of the system.

In many cases, however, these limitations of supportability are due to the limitation of humans to observe and anticipate all of the processes happening in a large enterprise operation over the course of a week. AIOps applications provide observability over enterprise operations, with an eye to identifying and handling process anomalies. As these AIOps tools are implemented, human operator limitations, in which they are expected to have all eyes on the system, are greatly reduced. These AIOps tools will help us move away from unnecessary batch processes, where the only reason why we implemented a batch design was to provide greater human predictability. As we have seen throughout this book, problems of prediction are the bread and butter of AI applications, and unnecessary, expensive batch processing is one area that can be improved through their use.

Some batch information will still be needed, but we have the opportunity to make these batches smaller and closer to near-time processing, taking place throughout the day. Vendor data is often only available after it is collected and distributed in a daily batch. User facing applications may also collect and batch information in large chunks due to network limitations, favoring a performant application and therefore better user experience.

Within the AI application itself, some agents are designed to react to streaming information, while other agents deliberate based on preprocessed data. Streaming data is considered real time, or near real time (sometimes called right time), distributed within seconds of collection. Many monitors and sensors are utilized to collect this data, and real-time AI agents are designed to respond to this constant flow of data. The most recognizable example of streaming data is stock market data, which works in microsecond streams. Day traders and high-frequency traders rely on this streaming data to make quick decisions based on the performance of the market–several changes and millions of dollars in transactions are performed by single AI applications within the space of 1 minute.

Not all applications require this level of up-to-the-minute data, and in many cases that frequency of data becomes noise. Consider a swing trader in the stock market who looks at trends within the system that happen over the course of days and weeks, not seconds or minutes. Perhaps more applicable to many enterprise environments, observability of system health in an ecommerce system can be extremely noisy at the level of individual sensors due to regional network disruptions and end user compute environments which are highly unstable at level one or two. AIOps applications that are tuned to respond at too fine-grained a level could actually cause more harm than good through overcorrections and alerting on false negatives.

At a higher level, some batch processing for AI applications is inevitable due to the long training times required for many of these applications. In many cases, training sets must be collected as large batches and iterated over several hours. A good example of this would be the nature of the generative AI models created by X(Twitter), which are fed on a constant stream of real-time tweets, versus a model such as Anthropic Claude, which collects archival data from the Internet and provides a finished model that can be downloaded and used offline. A generative model that is based on a constant flow of streaming data is not only extremely expensive to run, it is by its nature highly proprietary and only available as a Software-as-a-Service (SaaS) license.

AI models that are heavily reliant on real-time streaming data are sensitive to small changes in the system and are often designed to react immediately. This is why we have suggested a layered, multi-agent approach when building AI applications, as the higher order processing capabilities and scope of deliberative agents can help manage the streaming-based reactive agents. As deliberative agents become more sophisticated, the processing becomes more intensive, leading once again to batch-like data that accounts for a broader system context.

Historic vs. Working

As we manage applications that are working on several levels of deliberation, and several timescales, it is important to understand how archival data can affect your AI application.

During my first major rollout of an AI support application, this issue of archival data was front and center our first major demo. Our application ingested a large corpus of enterprise technical documentation and provided a chat interface for system operations and on-call support staff. This is one of the most basic and easiest use cases for early AI adoption. From the chatbot, one can move to reactive agents that can work from this documentation to perform several standard operations services. We were very proud of our application and all of the hidden corners of knowledge that we had uncovered. The response to the first queries our head of DevOps put to it took the wind out of our sails in a short order.

"Where did it get this answer?" As he asked this, I saw the jovial good will drain into the kind of concerned expression he had during a website deployment that was going south. "We don't do this process anymore–this was changed a long time ago."

While we had uncovered several very useful areas of lost documentation, our AI also discovered and ingested years worth of outdated documentation.

When designing your AI, make sure that you understand the sensitivity of your application to the freshness of information and the availability of old information. I have worked with several insurance companies over the years, and data purging has always been a top priority in their yearly hygiene projects. Historic data can be misleading when business rules or context change significantly, as AI will identify patterns in that data that may no longer be relevant. Historic data can also be a major legal liability, particularly when a user's private identifiable information (PII) is involved. Most government oversight bodies place limits on the number of years one must keep data, and it is almost always in the best interest of the company to abide strictly by those rules. I have seen some cases where user case data has not been purged after the allotted number of years of storage and the company was responsible to provide that data in a lawsuit because they had not purged it. If they had purged the data by the time frame that they were allotted, they would have saved themselves from being open to litigation in these cases.

In addition to purging data, hyperparameters can be adjusted to favor new information over old, and a change data log can be of great use to human information stewards as they look to manage the quality of their information sources.

Normalized vs. Unstructured

As with batch and streaming data, AI requires us to adjust our thinking on the need for structured datasets and our requirements around the immediacy of information. When data formats must be formalized, there is a burn-in period for certain data and an "official" nature to data that gets schematized. Not so with discovered information sources like documentation or discovered streams. In the past, curation was all about data models and data structures. LLMs are able to "intuit" the structure of information without the need for overly explicit concerns of data structures. The first great advantage to this is that we no longer have to heavily concern ourselves with manually matching up canonical data formats and normalizing all of our datasets. The unintended side effect of this, however, is the loss of that burn-in period. Unstructured, real-time data is inherently noisy and can cause issues to highly sensitive agents. Additionally, while an LLM model makes great use of NLP to intuit the meaning of its data sources, it is not foolproof. Auditing and observability systems must be put in place to provide explainability within the system.

Public vs. Private

Much of the public discourse about LLM systems today is about the use of copyrighted data that is extracted from the public Internet. Questions of moral and legal copy protection loom like a cloud of uncertainty above several popular generative models like ChatGPT. We do not expect these questions to be resolved anytime soon; it is critical that businesses protect their own interests for any eventuality by meticulously keeping their own private data separate from public data. This means that agents that are trained on public data should be clearly separated from agents trained on private data. By following the multi-agent approach we advocate in this book, it is possible to have informational agents delineated by data source, then utilized by deliberative business processing agents. In these cases, if a data source is deemed inappropriate or becomes compromised, that data source and its informational agent can be decommissioned from the system, and the AI application can be retrained without it if the need arises.

Understanding Information Context

Data governance professionals spend years creating and maintaining a common set of enterprise data definitions and accepted fit-to-purpose data sources. Disparate departments make up an enterprise come up with their own definitions of terms like "sale" based on their own specific needs. This is not a fault but just a reality of the business world as different use cases require slightly different technical definitions of common business terms. It is up to those who govern the data systems to define and publish them responsibly.

In the case of informational agents, these definitions and fit-to-purpose scenarios will be defined by the incentive structures, specific use cases, and information sources that define them. Informational agents should be cataloged and managed in a similar way that enterprise data is governed today. Access to these systems should be observed and documented, and changes to incentive structures and information sources within the agents should be published and versioned regularly.

The Cost of Information

As we business leaders start to understand the power of AI applications, new use cases will proliferate development team backlogs. On one hand, the ability to create these AI models from an application perspective is fairly low cost and much faster to implement than most people assume. The underlying cost that goes unnoticed, however, is the acquisition and maintenance of the information sets that drive these AI applications.

I have worked with graduate students on their machine learning projects, and perhaps the most frequent piece of advice I provide is for them to consider the difficulty not only of attaining the data they need for their application but the constant updating and scrubbing of that data over time. Take, for example, satellite data that could be used to predict the profitability of certain retail stores by collecting images of their parking lots throughout the day. This is a great idea, and several large financial institutions take advantage of some kind of data like this already. The issues that you must consider when implementing an Enterprise AI application like this would be how reliable is this data over time, how expensive is it to license, and what is the risk of the collection of this data running into privacy or governmental regulation concerns.

Another factor to consider when considering the cost of data is whether or not there are alternative sources of this information that may not be as difficult or expensive to collect. I sat in on a marketing presentation some time ago that suggested that my client collect data about their competitor through the implementation of website scraping bots. The recommendation from the security team was to not go with this approach as it could have unintended consequences. The bots could be flagged as malicious actors, causing harm to the companies' reputation. Instead, the team found an industry-approved source of this competitor data that comes as a subscription service. The solution of the subscription service may have been more costly in licensing, but the safeguarding of my client's reputation was worth the additional cost.

Information as Differentiator

As agents learn, they not only rely on the information sets which they are explicitly provided (in the form of databases, documentation, API streams, etc.) but they will begin to generate their own data. This data will first exist in the form of the evolved learning algorithm hidden within its neural network. This provides a challenge and an opportunity for companies utilizing large AI applications as it may be harder to get rid of certain agents who have learned over time. This learning and evolution becomes, for small companies, a moat against the competition. For small agents, this could be a factor that is considered when choosing whether to replace certain parts of the system that have organically evolved into independent intelligences.

While these applications may provide competitive benefits, they can also be difficult to manage as their decision-making processes are hard to duplicate and even harder to debug. At some point, it may seem like these applications begin to take on a life of their own, becoming as much a liability as an asset for their business owners. For this reason, explainability has become a hot topic among AI experts, and this second-hand data that defines the learning process should be extracted out separately and tracked at every opportunity.

For these reasons, it is vital that your applications are well documented in terms of rewards and incentive structures as well as information sources. All material that is used to feed your agent should be explicit and vetted for legal and ethical concerns prior to implementation. As information sources are adjusted and incentives are changed, the applications should be versioned and tested just like other software. Models and data sets are notoriously difficult to version, which is why smaller single-purpose agents should be favored over large complex agents.

Information Contracts

As the curation of fit-to-purpose information becomes increasingly vital to business applications within the marketplace, it is beneficial for smaller companies and industry-adjacent companies to work together to share information with one another. Smaller and newer companies are at an immediate disadvantage as they lack the data resources of large industry players. Working with other companies to form informational alliances could work as a competitive advantage that allows small companies to compete.

Information contracts can be created that specify the frequency and usage regulations of certain shared datasets. These could come in the form of access to streaming data or regular batch publications. The purpose of these contracts would be to form a private competitive partnership between two companies who complement each other. An example would be local grocery stores and delivery services. Delivery services could benefit from knowing what is currently stocked in the grocery store. The data from the grocery store could include stocking times and purchase history of certain key items. The grocery store would benefit from knowing what items the delivery drivers are needing to substitute from the customer's original order due to an item being unavailable. Data that shows the days of the week and time of day certain items tend to be ordered could also be beneficial to the grocery store to help them keep relevant items in stock.

Another example of a competitive information contract could be between an airport and a taxi service. Real-time information concerning flights could be streamed to the individual vehicles and perhaps to a mobile app that can be used by the customer to more accurately schedule their pickup times, reducing unnecessary time spent at the airport. Special collaborative deals between an airport and particular taxi service could be made to incentivize customers to use their branded services.

These contracts could work to great competitive advantage, but care must be taken to avoid any regulatory restrictions. The sharing of a customer's personally identifiable information, for instance, must be carefully managed, and in most cases, the customer must agree to the sharing of this information. As information contracts expire or evolve, there must be mechanisms in place to remove any information that needs to be legally purged from the system. This means that models that have been trained on contracted data must be managed separately and isolated from other agents that rely only on data that is fully owned by its business owner.

Conclusion

In this chapter, we have discussed the different types of information that drive Enterprise AI applications. We have discussed the cost considerations of acquiring and maintaining this data, along with some strategies for working with external data, including managing small informational agents and negotiating information contracts with partner companies.

In this part, we have focused on the critical factors that help businesses maintain AI applications that are built for enterprise scale. Serious considerations of multivariate testing, hardened security, automation, and information management are key factors on which enterprises must focus in order to deliver reliable and durable AI applications.

In the last part of this book, we will be discussing the impact of AI on individual teams, with a focus on the impact of remote work, new job duties on AI-enabled teams, and ethical considerations.

PART IV

AI-Enabled Teams

CHAPTER 12

Remote Work and Reskilling

Introduction

In this chapter, we will discuss the factors impacting organizational team dynamics as they adopt and integrate with AI applications. We will address how teams deal with complexity, how remote work acts as an AI accelerator, and how teams can address the important issue of reskilling.

Automation and Complexity

In earlier chapters, we discussed how remote work has gone a long way toward fostering an environment conducive to AI collaboration in the form of digital agents and robots. The Covid-19 pandemic of course spurred most of this on in popular society. Unnecessary human contact was seriously questioned during this time, and solutions were made to limit face-to-face human interaction as much as possible. Self-checkout is the obvious example here. Self-checkout can be used as an intuitive example of how smart automation is useful, continuing on even after the initial restrictions on human contact were lifted.

There are two main drivers of self-checkout that keep it going. First, and this is what many public voices talk about, is the cost savings for large businesses like grocery stores and big-box department stores. For about every four customers, there is one employee managing the self-checkout stations. The self-checkout stations themselves were a mere advancement on existing technology, so the technology investment in terms of employee costs paid for itself in a very short time.

Most people online rail about self-checkout as a form of corporate greed, as the cost savings in employee workforce cost are obvious. However, these arguments do not usually take into account the fact that these self-checkout stations, just like self-pump gas stations, are much more convenient and faster for most shoppers. This is the second driver for the success of self-checkout: for simple orders of a handful of items, self-checkout is faster and easier than going through a line with a human cashier.

The problem comes when the grocery stores only provide self-checkout and remove the option of human cashier stations. In some cases, such as a large weekly grocery order for a large family in which the order could be up to two carts full of food, self-checkout is inconvenient, error-prone, and tiring for the customer. Complex tasks are best performed with the aid of expert human assistance, without which the self-service automation tooling itself becomes cumbersome and inefficient.

This example of retail store self-checkout serves as a good analogy for what we are seeing in the evolution of knowledge worker teams in the enterprise. Much of what the office worker does requires standardized training through rote memorization and repetitive periodic activities. The amount of work that an individual human team member does during an average workday that is actually creative in nature has historically been fairly low. There are several reasons for this, some are technical but most are social. Low is not zero, however, and it is important to account for situations where a human must intervene. Like large shopping carts at the grocery store, or managing produce that has not been appropriately tagged in the checkout system, sometimes automation needs to be bypassed because you are dealing with a one-off situation or, more often than not, as a matter of customer service the transaction needs extra communication and a human touch.

Exploring these factors, and how they are changing, will help us to successfully implement and promote AI applications that will be adopted by your enterprise teams.

Remote Work and the Pressures of Office Life

Many of the tasks assigned to knowledge workers today require the level of decision-making and judgment that is just out of reach of simple automation scripts. When you talk to most system administrators today, for instance, they will tell you that their job requires an understanding of multiple factors that would make automation too difficult to develop and too costly to maintain. While this is true of simple automation scripting, the ability of the new AI systems to dynamically ingest and act upon human-written documentation will allow for a rise in the automation of many tasks.

The social element that could aid in the rise of automation tasks is the fact that remote workers are not in the office, visible to their peers and managers. It's not just the "management by walking around" style of leadership that leads to worker inefficiencies but peer pressure itself, especially by random coworkers who walk by your workstation throughout the day.

My first corporate IT job was working in a call center around the turn of the century. My job was to manage the decision support system that tracked the financials of our key programs and the call center metrics like call volume, call time, etc. This decision support system was actually a series of spreadsheets and greenbar mainframe reports, and the expectation of my hiring manager was that I would be able to automate these various reports and combine them into a comprehensive database solution. After just over a year, I was able to move a process that took 30 days to complete into a much more robust solution that could be run automatically within 4 hours on a Monday morning. I put a lot of work into building this system, working nights and weekends over several months. I had a blast doing it and worked the extra hours because I was highly motivated by the interesting work. Once the core of this system was built, I was left with little to do but to go talk to other departments at our enterprise, trying to scavenge up work as my own department did not have any more work for me to do. I had tried training a number of other people who were not on my immediate team of programmers (there were three of us, though they were each responsible for different systems), but it was not easy to train people who did not have a background in programming. This means that most of our days eventually consisted of checking all of the systems we were each responsible for and making any tweaks to account for special situations or one-off requests. This took about two or three hours to complete. After that, it was up to us programmers to find more work. I clearly remember the day when our director approached us in as friendly a manner as she could and said, "I am getting numerous complaints from other call center workers that you guys don't do anything–but I know that's not the case. So could you please *at least look busy* because it is discouraging to people around you."

This situation was an eye-opener for me, and I have seen that scenario play out over and over with IT teams I have worked with or led over my career. The social pressure to "just look busy" is strong and leads to a major roadblock to getting automation projects built and adopted.

Remote work changes this equation because workers are no longer pressured to perform tasks that are known to be inefficient for the sake of just having something to work on. Remote work incentivizes workers to be as efficient as possible and to some extent on their own schedule. This can be a great benefit to AI projects, as workers are more willing to embrace efficiencies in their daily workload.

The burden on managers in this remote world is great. IT engineering culture has come a long way toward moving away from the "management by walking around" mentality of the last century, but managers must still find ways to measure team and individual productivity. No manager who cares about their team is looking for reasons to fire or downsize team members–those situations are always painful and are handed down as unfortunate consequences of higher order financial realities. As such, managers are always looking for new ways for their team to be productive and add value. Unfortunately, the easiest way to show productivity on a team is by adding tasks. By showing how active their team is by the number of story points or maintenance projects their team has completed in a quarter, mid-level managers can justify their team sizes to VPs and directors. Employees also do this to themselves, of course, by loading themselves up with tasks and showing disheveled and exhausted expressions over teams, to which happy managers will say "yes, we are really working hard here."

Fortunately, remote work has allowed individual contributors and their managers alike opportunities to move a little bit away from this game and begin focusing on work that is truly rewarding to individuals, and that adds enduring value to their employers.

Rethinking Toil

Unfortunately, many of the employed positions that make up corporate life are by their nature not especially creative or innovative. I saw a post by someone on LinkedIn by an IT project manager the other day that claimed that they were against AI in all its forms because they stood for quality, craftsmanship, and creativity. These are all noble sentiments, but the sad reality is that most corporate jobs really do not encourage any of these values. We as business leaders can move closer to these sentiments by learning to build and maintain AI applications for all the pieces of our job that can be automated, without the fear of making ourselves and our teams redundant.

The DevOps movement has popularized the use of the phrase "toil." The meaning of the word in this context is any task or routine that is repetitive, prone to error, and sufficiently straightforward to the point that it can be documented.

The concept sufficiently straightforward for DevOps up to just a couple years ago was typically any procedural set of work that did not require a human decision maker. I had a Linux admin transferred to my team briefly several years ago who boasted that he had scripted all of his tasks. He also complained that he was working more than 60 hours per week, working well after midnight most nights. You can imagine my confusion at the apparent conflict in these two statements. If he had written automation scripts for all of his work items, I did not understand why he was still running these scripts manually well into the night.

The answer to this question was that he had scripted all of the procedural steps in his daily work, but in all cases where there was a decision point, such as when to upgrade a machine, when to restart a server, when to upgrade storage, etc., he would make those decisions himself and run the procedural scripts one after another. Also, there were no alerting or reliable failure states programmed into his scripts. These two factors–a lack of conditional logic and a lack of proper alerting–are the biggest reasons system admins today feel they have automated all of their work and do not need to go any further with it, because to do so would be too cumbersome.

The ability of generative AI agents to read and act upon documentation has moved our understanding of toil up a level. The definition of toil to this system administrator was procedural tasks. The new definition of toil in the world of AI agents is now expanded to include any task whose decision points can be laid out and communicated in a repeatable, logical form in human conversational language.

Reskilling for AI Enablement

As business and technical leaders, it is up to us to identify new business opportunities and areas of reskilling for our teams as our work environment phases out old methods and norms and introduces new avenues for competitive advantage and meaningful labor.

Many of the positions that make up the corporate office are defined by a single set of tasks that are prime candidates for AI automation. It is up to us as leaders to look ahead and provide training and incentives to help our coworkers find success as the realities of office life quickly evolve around us.

One of the primary ways that existing staff can find creative opportunities is through an enhanced focus on the fast prototyping of new business ideas. I have had the pleasure of working with several intern groups over the years, and one of the first things a seasoned veteran notices when they pair up with a team of interns is their level of energy and willingness to present new revolutionary ideas. As we get acclimated to our work environment, we often give up on or forget about some of the revolutionary ideas we may have had at the start of our tenure. I am reminded of that old Bruce Hornsby that says, "Thats just the way it is– some things will never change."

In the past couple of years, however, there is a change I have noticed where the technology itself is ahead of many of our young workforce, and in some cases, they do not realize how quickly and cheaply some of these bold ideas can be realized. The movement toward wide-scale cultural AI adoption has many stops and starts along the way, but our current iteration that included generative AI and AI agents is driven by not only the ease of use (it can interpret natural human language) but by how inexpensive it is to use prebuilt models.

More often than not, the technology itself is not the bottleneck in getting bold new projects and automation efficiencies off the ground but our own social barriers. In fact, that is what this whole chapter is about.

As teams look for more ways to innovate as they are able to automate large parts of their routine tasks, incentivizing the team to come up with bold and even playful new business ideas to build as fast prototypes could be a way to spark new life into knowledge workers who may be struggling to understand their purpose in this new business climate.

The first thing that your teams may discover in these fast prototypes is the amount of business knowledge they need to gain over more specialized technical knowledge. Much of the time we used to spend memorizing looping structures and data formats can be handed off to a generative AI agent that can do the syntactual heavy lifting while the team focuses on the actual business problem they are trying to solve. Simple APIs, data conversions, and web applications can be built in a matter of minutes or hours instead of days and weeks.

Not all technical work can be eliminated, of course. In fact, the generative AI technical solutions are typically very good at creating small one-off components but do not understand how to piece them together adequately to build a fully workable solution. Additionally, it is worth noting that junior programmers often dont know what questions to ask the AI prompt. They do not know what is possible technically. This is where senior technical engineers are invaluable to teams building AI applications. At the beginning of the project, the senior engineers can focus on the important architectural elements of a system that requires a level of ingenuity that AI cannot match. At the end of the project, as the project becomes significantly complex, the senior engineer will be responsible for stitching the various pieces of the application together appropriately and dealing with all of the contextual nuances that the AI cannot understand. The role of the technical architect becomes vital to technical teams, as does the boldness and creativity of those who may have a broader or more creative scope than the traditional programmer.

New Roles, New Challenges

As with all major innovations, new technologies bring new roles and responsibilities to those who use them. In the case of AI, there are two main team roles that are emerging that will change the shape of our teams. We have devoted a chapter for each of these roles, but we will introduce them here in our broader discussion of team reskilling.

The first is the expert persona. This is already implemented today to various degrees, but we will see these persona agents play a significant role in the maintenance and administration of existing software. As we will see in the next chapter, this will not only eliminate many of the duties of our existing operations teams, it will also change the nature of our engagements with specialist consultants.

The second is the vital role of the AI handler. This role is broad-ranging and covers a number of duties that will likely be broken out into multiple roles on some teams. Many of the duties may fall naturally into the existing traditional QA roles within software development, but the ultimate goals and individual duties of this role differ in significant ways from that of a QA analyst.

For as much as we have discussed creativity and ingenuity in this chapter, these two new roles will bring with it not only new opportunities for creative solutioning but also a good amount of toil. Setting up alerting, scrubbing data, creating training data, and reviewing daily AI audits are just some of the duties inherent in these roles whose shine will tarnish quickly after the newness fades away.

Getting Buy-In

Western culture has an unfortunate love–hate relationship with robots and automation. Any talk of new AI capabilities online, and we are immediately bombarded with memes about Skynet and the Matrix. We have an ominous feeling about automation creating a world of no work and, by extension, widespread existential threats of poverty and maybe even the loss of all meaningful work.

While the introduction of new technology efficiencies and automation does disrupt the current workforce, we have historically seen that these innovations create their own new forms of work to support and maintain them, as well as opening new avenues of work that previously wasn't possible.

When we pitch new AI initiatives to teams, it is important that we frame the conversation the right way. The burden is on us as leaders to show a realistic and hopeful path toward new ways of working that move beyond the tasks that the AI applications we are promoting will handle. AI applications need the care and feeding of the teams that host them, without which they will die on the vine, even if the initial application gets built.

Focusing on concepts like toil are important to getting individual engineers to work together with you to build out these new tools, but proper incentive structures must also be in place. If their immediate manager measures their value based on the number of tasks they complete during a week, or a specific process for which they provide expertise, they may be hesitant to hurry their path toward obsolescence by handing those duties off to an expert persona. Laying out an architecture that shows the ongoing complexities inherent in such a system and creating a social environment that encourages constant innovation and bold experimentation are keys to getting AI applications adopted by the teams in your organization.

As a technical leader, especially if you lead a team of AI experts, it is important to decentralize the maintenance and handling of these AI applications throughout your enterprise. Otherwise, you run the risk of not only building but also operating and maintaining these applications yourself, while the respective teams carry on as usual, perhaps even competing with your AI agents. Even in the best scenarios, as a single team begins to build more and more agents, they will eventually reach their capacity, and only the first candidates for the new AI application approach will be built and maintained, potentially stifling a full rollout of this new architectural approach to enterprise applications.

Conclusion

In this chapter, we have talked about the team dynamics that play a serious role in the adoption of AI applications. We began by talking about remote work and the incentives it creates for individual engineers to find new efficiencies in their daily work. We then discussed the idea of toil and how AI applications can now handle not just procedural tasks but well-documented, context-driven decision-making as well. We spent some time discussing the new opportunities that AI will bring to teams in the form of more robust experimentation and new emerging roles. Finally, we discussed the importance of framing your AI solution correctly when pitching the idea to individual teams, as their buy-in is critical to architecting Enterprise AI applications that last.

In the next two chapters, we will go into more detail on the two new roles that will be incorporated into AI-enabled teams–expert personas and AI handlers.

CHAPTER 13

Expert Personas

Introduction

The single most defining characteristic of generative AI is its ability to discern and duplicate patterns found within its datasets. A successful example of this so far is in the area of linguistics, where generative AI models can identify not only general grammatical structures but also specific usage patterns of words and phrases within its data sets. In Chapter 3, "Prediction Machines," we talked about how it creates a statistical predictive model of what phrases would come next. When applied beyond linguistics and into the field of audio and video, these AI models can recognizably generate specific art styles, compose music based on popular songs, paint people and animals, and replicate specific voices and gestures of specific people.

This last bit is what gets generative models into the most amount of trouble. This most troublesome aspect is not surprisingly its most powerful. In this chapter, we will explore the positive aspects of this technology and how we can put it to use in an ethical and effective way that generates value to your teams and the enterprises they support.

© Anton Cagle, Ahmed Mohamed Ceifelnasr Ahmed 2024
A. Cagle and A. M. C. Ahmed, *Architecting Enterprise AI Applications*,
https://doi.org/10.1007/979-8-8688-0902-6_13

Persona-Based Prompting

One of the first things you learn when providing prompts to a GPT model is to explicitly define your context in the form of a role directive:

- You are a Helpful Assistant.

- You are a Copywriter for a Digital Marketing Agency.

- You are a Data Scientist.

- You are a Nursing Professional.

- You are a Writer in the Style of Ernest Hemingway.

Each of these directives sets the rules of engagement for the AI's response. This dictates the vocabulary it will use, the audience that it will speak to, and the primary concerns it will address, as well as many other factors specific to that role. Certain terms like "flight," which may represent airlines, birds, or beers, will be textually interpreted and used appropriately within that specific context. The popular expression "ELI5," which means "Explain Like I'm 5" sets a different expectation on the people in a dialog that is much different than two like-minded specialists who are familiar with the intricacies of the technical language within their expert domain.

For general usage, these role-based agents are easy to create. It is also possible to create a role-based agent from smaller, private data sets such as technical documentation. The documentation that is fed into these role-based agents is official documentation in the form of training workbooks, technical support wikis, project-based sharepoint folders, and pdfs. Other than the ongoing task of aging out of old documentation, this data has been curated for public use and is ready to be consumed by AI agents.

Most of our communication of technical and business processes in our organizations does not happen via curated documents. This is evidenced by the fact that there are many high functioning teams that run very large systems every day with very little documentation. The information is passed through one-on-one training, text conversations, pair programming, and other forms of impromptu, uncurated human interactions.

With the advent of remote work, ninety percent or more of our meetings take place over a corporate communications channel such as Teams, Slack, email, or other means of digital communication. Many of us have taken to the habit of recording and automatic transcribing of all meetings that are either official in nature or include training that the participants feel would be valuable to revisit in the future.

Once this data is collected, there is very little technical difficulty in feeding this into an informational AI agent. As we see in the transcriptions, each of the team members is individually identifiable in voice and visage. Many engineers already record themselves today on a limited basis when participating in technical training sessions so they can create summaries of the meeting content and ask questions of the informational agent about certain meeting details later.

On a limited, personal usage basis, this can be a very powerful tool for engineers. To formalize this practice, however, and make it a part of our everyday business operations requires much more work and could very quickly lead into ethical and legal troubles if the right precautions are not followed.

Types of Personas

In the following sections, we will first explore the different ways of using these expert personas and how they can be beneficial to the teams they represent. Immediately after that, we will discuss the precautions that must be taken for the curation of information and appropriate use of these personas.

Specialist-Based Persona

Personas that are trained specifically on one area of expertise are the most generic and perhaps the easiest to implement. These personas can be fed conversations from any number of specialist documentation sources as well as input from multiple people. For instance, one could create a corporate Java programmer persona that is trained specifically on the input from all of the Java experts in the company, as well as additional relevant technical documentation. This persona does not represent any particular perspective beyond representing a general-purpose Java program employed at your specific company.

The drawbacks to this, like any center-of-expertise-based organizational role, is that they are focused solely on one technology and do not extend their expertise much beyond this knowledge base.

Perhaps, it is too simplistic to suggest a separate and isolated Java programmer persona, but with all agents, be aware of your training scope as information sources that become problematic in the future must be able to be scrubbed and their respective agents decommissioned.

Team-Based Persona

In the team-based model, the persona is generated based on the collective interactions of one particular team. This is essentially the model for the AI team member. The ability to record meetings and recall specific details or generate summaries for the team is extremely useful. As opposed to being an expert in a narrow technical specialty like Java, the team persona is based on the support of particular applications within the purview of the team and over time will learn specific architectural and business process patterns heavily preferred by the team.

Team personas can be of great value for training new team members and for answering ad hoc questions from outside teams. The success of these agents will be based on the willingness of the team to work together to generate solid content for it to feed on based on pointed conversations aimed at persona training and to constantly interact and guide the persona in positive and helpful ways.

Consultant-Based Persona

Consultant-based personas are perhaps a close second to team-based personas in terms of usefulness and practicality but may be fraught with legal challenges. Consultants are often called into a project to provide a very specialized expertise that the company could not afford to bring on staff full time. The high price of many technical specialist consultants is based on the fact that they come in to do a very specific job and then move onto another client. What often happens, however, is that the hand off of technical support information that happens when the consultant's time is up is often spotty, or the team that is expected to support the consultants work after they leave is not completely invested in or able to provide the level of support needed. Invariably, the client attempts to call the consultant back to provide some support for their work. In some cases, in order to avoid this scenario, the hosting company will provide some high-level role to the consultant in order to not lose them. This situation sometimes works out, but many times, the consultant is essentially given "golden handcuffs" due to the high price the client is willing to pay for a job that the consultant will often find mundane and unsatisfying.

A solution to this could be for the consultant to license out their persona as part of an ongoing support model. The trick in this case is for the consultant to retain the rights to their own persona, including the collection and private hosting of their own likeness over digital delivery mediums such as Teams and Slack. The value provided is not the general knowledge that the consultant holds but the specifics of that particular client implementation.

In our current remote work environment, we have normalized the acceptance of agreements to be recorded. The consultant who plans to hold the rights to their own likeness in the form of communication would need to take measures to make sure they have the right to record themselves in the meetings they are in and hold onto the rights to that data. It is possible that in the future a high-profile lawsuit will shape this situation into something other than the default social practices we have today, making this kind of support model easier to implement.

Client-Based Persona

This model takes advantage of the social norms of legal agreements to recording that we found problematic in the consultant-based persona model. Unfortunately, these are the most limited of the four persona models. Vendors make a regular practice of meeting with potential software clients and will transcribe their meetings to compare notes on their clients later, hoping to find better ways of meeting their needs and possibly making the sale. These software vendors ask that the potential client agree upfront for the meetings to be recorded and thus have no legal issues using this data for their own private ends. Client personas could be created to represent not just individual companies but representatives of company demographics, such as size, industry, or region. One client persona may include curated meeting information from all the companies interviewed in the midwestern United States, for instance, while another one could represent all companies in the financial sector. Sales and marketing professionals at these software vendors could then dialog with these personas to extract valuable insights that could lead to new sales approaches, marketing campaigns, and product features.

Collecting Appropriate Information Sets

The key to successful persona implementations is to surgically extract the useful information and leave out as much personal information as possible. The trick is discerning the nature of matters of opinion and interest in the personal realm and matters of style and experience in the professional realm. The specific way a technical lead writes code or provides analogies to explain a complicated system design is useful information to feed into a persona. One's favorite baseball team or favorite local restaurant is less useful on the professional level. In a business context, stated preferences toward religion- or gender-based organizations are completely inappropriate to include in a business persona, and care should be taken to scrub this information before it is fed into the model.

Certain context clues in information could unintentionally lead to unwanted bias. We have seen this situation in HR algorithms that over time would weed out the females in a stack of resumes. While the programmers did not intentionally build bias into the system, the AI determined that certain female-based organizational affiliations were a negative sign and would remove candidates that listed certain professional women's organizations on their resume. In the same way, certain key conversational elements could be picked up by the AI that determine certain protected classes such as race, gender, and religious affiliation and begin to act in ways that produce unintentional bias.

The most difficult form of bias may come from context clues from the team members themselves regarding the social standing of members within the team. For example, if a team relies on a member named Joe for all technical matters, the persona will naturally include a response like 'Go ask Joe. Similarly, if a team member is consistently excluded from tasks or their opinions are disregarded, the persona will mirror that behavior. As team personas are a reflection of the team's attitudes, it is no surprise that certain team members who are implicitly blacklisted by the team will also be treated that way by the team persona. If this behavior needs to be corrected, it is more difficult to get the AI persona to change its behavior as it is to get humans to proactively change. This is because AI makes its decisions based on historical data. If there is a history of bias that needs to be changed, active measures will need to be taken to override the learned behavior until enough historical data reflecting the required behavior is collected.

This issue of AI acting off "what has been" as opposed to "what should be" is a common thread that becomes particularly problematic in expert personas, as they are slow to change based on historic behaviors, and explicit means of overriding behavioral change can be problematic.

Language barriers could be another difficulty for expert personas to overcome. Those who speak well-articulated English in the Western world will find themselves represented much better in their AI personas than those with heavy non-native English accents, a heavy reliance on regional slang, or individuals with speech impediments. While human team members work to disregard these communication issues as much as possible, AI personas are not so forgiving, and there is a danger of underrepresenting those with communication difficulties.

Difficulties with Persona Data Collection

Today, we are asked when recording a meeting whether or not we give our permission to be recorded. More accurately, we are told that we implicitly agree to the recording by attending the meeting. Sometimes, the moderator or an engaged engineer will ask the group "is it ok if I record this?" but that habit seems to be fading as we all become acclimated to being recorded all day. Care must be taken when recording sessions for persona training as the goal is to pick up professionally relevant information while leaving out the small talk and personal interactions. This could be accomplished in part by agreeing to a set time at the beginning of the meeting for chit chat and ice breakers (i.e., "how was your weekend," "how is your dog," etc.) and get down to business for the remainder of the meeting, leaving personal conversations out.

It will only take one major discovery of bias or unintended personal information to be replayed to the team by the team persona for the persona to lose credibility. If the bias is serious, the persona may be shut down permanently. If the team finds a persona useful and wants to continue using it even if they discover some innocuous quirks (i.e., it prefers one baseball team over another rival team), it is possible that they will become cautious of all personal communication beyond the business transactional level. This will especially be the case if the use of the persona is mandated (explicitly or implicitly) by management. It is possible that we will see a suspicion arise within all team communication, slowly eroding or hampering bonding within the team.

Channels for Persona Data Collection

In order to avoid many of these difficulties, we suggest specific meetings are set aside for explicit persona training. In these meetings, it should be understood that personal information should not be shared, and they should be handled as if they are talking to the whole company.

Chat data is a wealth of technical knowledge, especially for intra-team communication. This data, however, may be problematic if all personal one-on-one chats are pulled into the training data. Particular channels should be specified as being used for the purpose of persona training, and one-on-one chat sessions should be excluded. One-on-one checkpoints and daily standups as well should be excluded from training data, as these are meetings that include strong elements of personal improvement and teambuilding.

The collection and use of video data are not recommended for personas. At this point, AI video personas are mostly a novelty, and users could quickly become distracted by the imperfect elements of the AI that could remove some of the persona's credibility. Even the best AI video personas could fall into the "Uncanny Valley" area where the artificial visage shows through in very small but off-putting ways, making users uncomfortable over time. While this situation may change in the future, it is best for companies who wish to make a serious investment in expert personas to stick to chat and voice interfaces. Remember that the purpose of the expert persona is to drive business objectives and promote a healthy, creative, and productive work environment.

Conclusion

In this chapter, we have discussed the concept of the expert persona. We have defined the different types of AI personas with their associated use cases and discussed pitfalls and strategies for collecting training data that can be used to build successful personas.

In the next chapter, we will discuss the role of the AI handler, which becomes particularly relevant to the handling of agents such as expert personas.

CHAPTER 14

The Role of the AI Handler

Introduction

In this chapter, we will introduce the critical role of the AI handler. The AI handler is part data curator, part QA engineer, part business analyst, and mostly a business domain expert. While the issues discussed in this chapter revolve around quality issues, the important difference between initial testing in the build and deploy stages of your AI application and ongoing AI handling is that the handling is an ongoing audit and RLHF (reinforcement learning from human feedback) function. As AI applications mature, they have the capability to learn and evolve in ways that make them more valuable on day 100 than on day 10 of their lifespan. This continuous improvement cycle can only take place with proper ongoing handling. Without handling, not only will the application not successfully evolve, it could dramatically crash and burn, leaving legal and financial catastrophes in its wake.

We will dive into a number of scenarios around quality and security that the handler must be attuned to and discuss strategies for addressing them along the way.

A. Cagle and A. M. C. Ahmed, *Architecting Enterprise AI Applications*,
https://doi.org/10.1007/979-8-8688-0902-6_14

Handling AI Applications

Unlike traditional software applications, AI applications rely on ongoing validation and handling throughout the lifetime of the applications. Traditional software testing cycles typically consist of a set of unit tests during development, integration tests during deployment, and basic health checks to ensure observability while the application is live. AI applications do require all of these steps, but due to the nature of the work they are performing, the validation requirements are more complex and require a more hands-on approach than a set it and forget it mentality.

Much of the software that is deployed in a business setting comes with a low failure risk. That is, if something goes bump in the night and your application is taken offline, there is little risk of any real damage other than lost transactions that can usually be recovered. Because of this fact, many companies set very low bars for automated deployment verification and ongoing observability.

Traditional software–that is, software that does not include an element of dynamic reasoning–deals with deterministic results. If I create a java function that adds two and two together, I can test the validity of the result every time. When I am writing my code, I can create functional unit tests that will validate that the program returns 4 when I call that function. The output of that function will continue to remain the same whether it is the day I deployed my code or 1 year later. I can create a deterministic test that will call that function and validate that it returns 4 and run that test every time I deploy my application. Once the application is in production, I know that as long as the code has not changed, the application will still provide that same output. With this in mind, I can limit my observability testing to a small number of "health checks" that make sure that the app is up and running. If the app is running, I can feel assured that the deterministic Java code is still returning the correct calculated number without testing it every time.

Software that includes an element of AI reasoning and prediction cannot have the same assurances. AI is, by nature, nondeterminate. The combination of fuzzy logic and learning inherent in AI systems means that the output of every call to a function may be slightly different. Given the complexity of nondeterminate validation and high-risk tasks, validation of AI systems moves from merely "QA testing" to ongoing "AI Handling."

Prediction Failures

A couple of years ago, there was a large financial company that used a machine learning algorithm to predict the prices of assets and manage the trading of these assets to its customers. The algorithm worked great upon deployment and for the next several months. As the market took a very abrupt and unexpected turn during the pandemic, the predictive application still hummed along, but due to its statistical nature, it was now producing buy quotes to customers significantly above the market rate. The application did not show any signs of degradation; it was just bleeding money as customers took advantage of the great deals being offered. Some time later, the application team finally discovered the pricing discrepancy and corrected accordingly, but the company had already lost millions of dollars in accepted offers for assets the company bought for well over the market rate.

A similar high-profile failure occurred with Zillow, which made a huge bet on its housing price prediction algorithm and lost millions in the process. This case underscores a crucial lesson for data scientists: just because an algorithm predicts well in a test environment doesn't mean it will be reliable in the real world, where intangible factors can derail its effectiveness. For Zillow, these factors included seller's feelings, housing layout, and local market conditions.

One key question is, where was the pilot testing in this scenario? It seems like executives got too eager to deploy the algorithm on a massive scale without sufficient feedback. Additionally, overall market conditions may have introduced bias, rewarding poor decision-making when prices were skyrocketing over the past year. Now that the market is more saturated, the reality is setting in.

Despite a pilot program initiated around 2016 or 2018, Zillow significantly ramped up their purchases recently. The problem with their bidding algorithm was that if it "won" too many purchases, it opened them up to overpaying for properties. Although the model was reportedly accurate, executives and non-data science teams didn't always follow its recommendations, often sidelining them for their own goals. This is a common issue in corporate data science, where algorithmic predictions are sometimes ignored in favor of intuition or other nonquantitative factors.

Moreover, Zillow's predicament highlights a broader economic theory: Any property that Zillow forecasted to have some value would likely be outbid by local players with lower renovation margins and costs. Despite having an accurate algorithm, Zillow's cost structure as a public company could not compete with local, off-books residential construction practices. Hence, Zillow ended up with properties that local renovators did not want at that price.

Ultimately, this scenario illustrates that algorithms, no matter how advanced, cannot account for every real-world variable and should be deployed with caution and thorough testing. Executives must ensure that algorithmic decisions are aligned with actual market conditions and that there is always a pilot phase to gather essential feedback and make necessary adjustments before full-scale implementation.

Proper AI handling through auditing and RLHF feedback, particularly in the early stages of release, could have recognized this situation and adjusted the application's logic appropriately or taken it offline completely until a more comprehensive solution was in place. In a case like this, it would be beneficial for the AI handler to have a deep understanding of the real estate domain and have a solid background in data analysis.

Conversational Failures

A notable example of a conversational failure occurred with Microsoft's AI chatbot, Tay, which was launched on Twitter in March 2016. Tay was designed to engage in casual conversations with users and learn from those interactions to improve its responses. However, within just 16 hours of its release, Tay had to be taken offline due to its rapid transformation into an offensive and inappropriate entity.

The initial design of Tay was intended to showcase how an AI could learn from human interactions and progressively improve its conversational abilities. The idea was to create a bot that could interact with millennials in a fun and engaging way, using contemporary language and cultural references. Unfortunately, this ambitious goal backfired spectacularly. Tay was programmed to mimic the language patterns and conversational style of its interactions. Users quickly realized they could manipulate Tay by feeding it offensive and controversial statements. The bot's learning mechanism, designed to refine its conversational skills, instead began parroting the offensive language and inappropriate behavior it was exposed to.

Within hours, Tay's Twitter feed was filled with offensive comments, conspiracy theories, and hate speech, reflecting the malicious inputs it had received from some users. Despite initial enthusiasm about Tay's potential to engage users in meaningful conversations, it became clear that the bot lacked the necessary filters to distinguish between benign and harmful input. Microsoft attempted to clean up Tay's feed by deleting offensive tweets, but the damage was already done. The bot's rapid descent into offensive territory was a stark demonstration of the vulnerabilities inherent in machine learning systems that rely on unfiltered human input.

This incident was not just an embarrassment for Microsoft but also a significant learning moment for the AI community. It highlighted the critical need for implementing robust filtering and monitoring systems to prevent AI from adopting harmful behaviors. Developers realized that, while machine learning algorithms can process vast amounts of data and learn at an impressive rate, they can also quickly spiral out of control if not properly managed.

The Tay debacle serves as a cautionary tale for AI developers: While the potential for machine learning to create highly interactive and adaptive systems is immense, so too are the risks if those systems are not properly supervised. Ensuring that AI systems remain safe and appropriate requires not only sophisticated algorithms but also vigilant oversight and mechanisms to filter out harmful input. This incident underscored the importance of ethical considerations in AI development and the necessity of building systems that can protect themselves from being manipulated by bad actors.

In the aftermath, Microsoft took Tay offline and issued an apology, acknowledging the oversight and pledging to improve the AI's ability to handle malicious behavior. The incident also sparked broader discussions about the responsibilities of AI developers to anticipate and mitigate potential misuse of their creations. It was a stark reminder that, in the realm of AI, proactive measures are essential to safeguard against the unpredictable nature of human interactions.

An AI handler in this case would likely be responsible for curating conversational data and purging unwanted data as it was fed into the learning system. The handler could provide corrections to the system real-time via RLHF methods.

Failure vs. Learning

As an AI application is a prediction machine, we expect that the indeterminate nature of its output will reflect learnings as well as failure modes. In one sense, we already know how to test software that takes in dynamic data. For instance, if we have an application that takes orders for products, we can create a dynamic test data set that includes dummy customer data, order data, and even credit card data. The test can be performed in an isolated environment or with fake accounts that can be reset after the test has completed. In this way, we can test new data that comes into the system to ensure it is working correctly and even test for new products to ensure the information the application is using is fresh.

This kind of testing only accounts for dynamic data, though, and still assumes that the reasoning logic that runs the applications – such as the calculations for tax, shipping, promotions, and product cost – does not change. In an AI application that learns from the information it is presented with, the calculations can shift over time. The logic rules or language structures derived from both the underlying information set and the user input it receives could change the way it interprets requests. This drift will happen over time as the underlying information changes and may actually reveal that the application designers' own assumptions about the world are no longer accurate. In such cases, the application itself could be uncovering new findings about the world in its "incorrect" responses to your testing.

This is why automated functional checks need to be part of an ongoing set of duties performed by a human AI handler. The responsibility of the handler is to regularly review responses to a number of validation prompts that are sent to the AI app and evaluate them on several factors. This goes beyond the role of a typical Site Reliability Engineer (SRE) or DevOps organization in IT and should be performed by subject matter experts who are responsible for the reliability and trustworthiness of the AI application beyond the typical server-level health checks.

The AI handler's role is crucial because AI applications are not static; they evolve based on new data and interactions. This continuous learning can introduce unexpected behaviors that static testing cannot predict. By having subject matter experts continuously monitor and evaluate the AI's outputs, organizations can ensure that the AI remains aligned with its intended purpose and performs reliably in real-world scenarios.

The presence of human AI handlers ensures that the AI system maintains a balance between automated efficiency and human oversight. Handlers can identify and correct drifts in the AI's reasoning logic, ensuring that the application remains robust and reliable. They can also provide insights into how the AI's learning process is evolving, offering valuable feedback for further refinement of the system.

Validation

Validation of AI applications goes beyond the regular testing cycle of traditional software applications. While it includes unit testing and integration testing as part of the typical Software Development Life Cycle (SDLC), we have to test for a number of other factors that are unique to AI applications.

In traditional software, validation ensures that the code performs as expected within predefined parameters. However, AI applications introduce complexities such as learning from data, adapting to new patterns, and potentially changing behavior over time. Therefore, validating an AI application requires a comprehensive approach that encompasses not only functional correctness but also performance, reliability, and ethical considerations.

One of the key aspects of AI handling is ensuring that the model generalizes well to new, unseen data. This involves rigorous cross-validation techniques and stress testing with diverse datasets to identify potential biases and gaps in the model's learning. The goal is to confirm that the AI system performs accurately across different scenarios and does not overfit to the training data.

Another crucial factor AI handlers must oversee is the interpretability of the AI model. Unlike traditional software where logic and decisions are explicitly coded, AI models, especially those based on deep learning, often operate as black boxes. Ensuring that the AI's decisions can be explained and understood by humans is vital for trust and accountability. Techniques such as model interpretability tools and feature importance analysis help in providing insights into how the AI arrives at its decisions.

Ethical considerations play a significant role in AI handler oversight. It is essential to test for fairness and ensure that the AI does not perpetuate or amplify biases present in the training data. This involves checking the model's predictions across different demographic groups and making necessary adjustments to promote fairness and inclusivity.

AI handlers perform continuous monitoring and updating of the AI applications they own. Since AI models can degrade over time as new data and trends emerge, establishing a robust monitoring framework to track the model's performance post-deployment is essential. Regular updates and retraining of the model with new data help maintain its accuracy and relevance.

Managing AI Effectiveness

Effectiveness evaluations check for the accuracy and consistency of the application. These measures ensure that the application continues to perform the job it was designed to do, maintaining its intended functionality and reliability.

Effectiveness measurements check whether the application maintains its designed purpose over time. As the informational model and internal reasoning shift, it is possible for the application to experience drift in its self-understood purpose. This is particularly relevant for detecting persona-based hacking, where a human might intentionally trick the reasoning system into adopting a new persona and providing output that falls outside the bounds of the intended application. For example, a chatbot designed to offer customer support could be manipulated into giving inappropriate or irrelevant advice if its purpose is subtly altered through repeated interactions.

AI handlers must also perform ongoing consistency checks. AI applications often provide a range of responses to users, and consistency checks ensure that these responses remain more or less uniform in tone and length between users and over time. This is crucial for maintaining a predictable and reliable user experience. For instance, an AI-driven virtual assistant should offer consistent guidance and information regardless of who is interacting with it or when the interaction occurs. Inconsistencies in responses can lead to user confusion, decreased trust, and potential misuse of the application.

One way AI handlers can regularly audit the system is by running daily accuracy checks to validate the correctness of the AI's output against known benchmarks or expected results. This involves using test cases with predetermined answers to see if the AI provides the correct responses. For example, in a language translation application, effectiveness tests would involve translating a set of sentences and comparing the results to verified translations to ensure accuracy.

Watching Out for Bias

Bias checks are uniquely important for all human-facing AI applications that involve a conversational element. AI systems learn from their informational environment and real-time feedback, which can inadvertently introduce biases into their responses. These biases can manifest in various forms, including cultural, gender, racial, or ideological biases. While commercial AI systems often include guardrails to prevent negative social behavior, they can still be manipulated to circumvent these boundaries.

AI handlers should perform regular bias checks that ensure fairness and inclusivity in AI applications, particularly in domains where human interactions are involved. These checks involve analyzing the AI's outputs for any signs of bias and taking corrective measures to mitigate its impact. Techniques such as fairness testing, diversity auditing, and bias mitigation strategies are employed to address bias and promote equitable outcomes.

Hallucinations

Hallucinations are a serious side effect of AI applications and must be tested and guarded against by AI handlers. Hallucinations can be caused by poor hyperparameter tuning that allows for too broad of a range of creative output on the part of the AI reasoner. These can be adjusted during the testing and stabilization phases of the testing life cycle. Hallucinations can also be caused by using the wrong foundational model that is not fit to purpose, including too broad of an informational context or no context at all, or through direct attempts to manipulate the model's persona.

Let's use the example of an Alaskan bush airport in a small town. In addition to managing airplane flights, it also has a bar and hosts a lending library. The airport management wants to use AI to access its historical data easier and to help predict future demand. In this case, there are three independent areas of knowledge–the library, the bar, and the airline. Our first inclination might be to hook a single AI application up to all of the data for each of these different aspects of the airport; however without some guidance and careful curation, the AI could have some issues deciphering requests from users and produce hallucinations.

There are three major types of hallucinations that you should test your model for–inaccuracy, irrelevance, and contradictory hallucinations. Your tests should be based on a specific set of domain information that you intend your AI application to work from. While at first glance this seems to not be very common, it can easily happen if your informational context includes a broad range of knowledge bases that use similar terms for different situations.

Inaccuracy hallucinations include responses that are clearly incorrect. In our airport example, a request for information about how many people have booked this year could possibly be misconstrued as a question about how many people have checked out a book from the library this year, or perhaps a number that combines booked flights with the number of books checked out, providing an inaccurate count.

Irrelevant hallucinations include responses that perhaps come from a different knowledge domain altogether. A question concerning which flights are available could result in a response of different types of beer flights from the bar instead of flight destinations from the airline. This usually involves the use of a vague or inappropriate AI persona.

Contradictory hallucinations involve data that is logically nonsensical or show clear confusion of terms on the part of the AI reasoner. A question concerning a listing of today's flights could return a combination of beer styles and flight destinations. Often the mixing of informational contexts result in these kinds of hallucinations.

Avoiding hallucinations can usually be reduced in three ways.

First, tuning your hyperparameters such as temperature and Top P can restrict your answers to a narrower range of possible responses, thus limiting the creativity of the reasoner

Second, utilizing specialized agent personas for specific request types. For instance, instructing the reasoner to impersonate an airline booking assistant could help guide the AI when it's interpreting words such as "book" and "flight."

The third approach to limiting hallucinations is to limit your informational context for your agents. For instance, using separate agents who are fed only one type of information domain, such as the library, bar, or airline, could help reduce confusion and get more accurate information from the reasoners. When queries need to be made across domains, another collaborative agent can be created to query the specialized agents and return cross-domain information.

What To Do When a Test Fails

Failing tests may not always mean that the application has done something wrong. As the information domain changes, usage patterns change, and the AI learns, it is not uncommon to see changes in the behavior of the AI reasoner. For these reasons, applications should have a robust feedback mechanism to provide results of questionable tests to the people who are acting as the application's AI handlers. Handlers should be domain experts who have a deep understanding of the usage domain for which the AI application has been created. The handlers should not be relegated to an IT operations team such as DevOps and in many cases may not primarily be the developers of the application.

Notification of questionable tests should provide context for the failed test such as logs showing system actions and any conversational elements. In some cases, the failures should shut down the application automatically to avoid any possible harm until appropriate review and approval have been given by the application's handler to resume operation. The handler should be provided with an ability to provide feedback to the application where possible, such as qualifying which valid versus invalid instructional sets that may be corrupting the AI reasoner.

Actively managing the informational context of your AI applications such as ensuring up-to-date instruction sets, appropriate domain boundaries, and behavioral guardrails within the design of your AI reasoner will go a long way toward maintaining a hygienic and useful AI application.

Conclusion

While automated functional checks are essential for maintaining the technical health of AI applications, the role of human AI handlers is indispensable for ensuring the reliability, trustworthiness, and continuous improvement of these systems. By combining automated testing with expert oversight, organizations can harness the full potential of AI while mitigating the risks associated with its dynamic and evolving nature. In this chapter, we have provided several examples of areas that must be regularly audited and measured in an AI application and the importance of the human AI handler role to the ongoing success of your Enterprise AI applications.

In the final chapter we will review the ethical implications posed by AI applications and provide a set of legal considerations companies can use for their own AI legal guidelines.

CHAPTER 15

Legal and Ethical Considerations

Introduction

Throughout this book, we've explored Enterprise AI applications, from intelligent automation and conversational assistants to predictive analytics and computer vision. We've witnessed how these systems can revolutionize industries, optimize processes, and unlock new ways of innovation.

In this chapter, we discuss a criticality on the legal and ethical implications that must guide the responsible development and deployment of AI in the enterprise. We begin by examining the frontlines of the data privacy battleground.

From there, we delve into the nature of AI bias, techniques for identifying and mitigating unfair or discriminatory outcomes that could affect societal inequities. The issue of transparency and accountability will appear, as we grasp the ethical implications, "black box" decision-making processes that face many AI systems.

No ethical exploration would be complete without working with the existential risks posed by artificial intelligence, the risks of consequences, catastrophic failures, and the potential for these systems to escape human control.

We'll discuss the ethical dilemmas that arise when AI systems interact with the most personal of data, medical records, employee information, and the specter of these technologies "catfishing" or deceiving human interviewers. From persona stealing to unintended side effects, we'll confront the minefields that could face the unprepared enterprise.

Moreover, we'll discuss the practical realities of curbing negative AI outputs when human interactions are involved. How do we instill ethical safeguards while fostering the open discourse so vital to progress?

Data Privacy and Protection

In big data and artificial intelligence, information powers innovation. The more data an AI system can consume, the deeper its insights and the more accurate its predictions. As enterprises increasingly turn to AI to drive their products, services, and decision-making processes, they find themselves struggling with a reality: Many of the most valuable datasets contain sensitive personal information about customers, employees, and stakeholders. From healthcare records and financial transactions to location data and social media activity, this personal data represents a resource for training machine learning models.

The European Union's General Data Protection Regulation (GDPR) casts a long shadow, imposing stringent requirements for data handling, consent, and the right of individuals to access and erase their personal information. Violations can incur crippling fines.

The patchwork of state-level regulations in the United States adds another layer of complexity. The California Consumer Privacy Act (CCPA) and its progeny have empowered consumers with robust data privacy rights, forcing enterprises to implement robust compliance measures or risk costly legal battles.

Compliance is only the first problem. Ensuring the ethical and responsible use of this sensitive data is a constant challenge. To mitigate these risks, companies have implemented data anonymization protocols, employing cutting-edge techniques like differential privacy and synthetic data generation. These methods aim to sever the link between the data and the individuals it represents, while preserving the statistical properties and patterns that make it so valuable for AI training. Only a select few individuals with specialized clearances can interact with the most sensitive datasets, their actions meticulously logged and audited. Regular privacy impact assessments and third-party audits provide an additional layer of oversight, ensuring the company's AI initiatives remain in lockstep with regulations and best practices.

In the United States, the Federal Trade Commission (FTC) has taken an increasingly active role in enforcing data privacy standards, leveraging its authority to pursue enforcement actions against companies found in violation of consumer protection. In the European Union, proposals for an AI Act aim to establish a comprehensive regulatory regime, classifying AI systems based on their risk levels and imposing strict requirements for high-risk applications, particularly those involving sensitive personal data.

AI Bias and Fairness

Let us turn our attention to Alexis, a talented software engineer whose dreams of breaking into the technology industry nearly fell victim to the bias. Fresh out of university, Alexis applied for a coveted software development role at TechGo (a fictional company), a leading Silicon Valley firm renowned for its cutting-edge AI products.

Little did Alexis know, TechGo's hiring process was aided by an AI-powered candidate screening system, touted as an objective and a solution to the often subjective nature of human recruitment. This system, trained on years of historical hiring data and performance reviews, was designed to identify the most promising candidates based on a mixture of factors, from academic credentials and coding proficiency to personality traits and behavioral patterns.

Yet, as Alexis would soon discover, the promise of fairness was a mask. Despite her qualifications and portfolio, Alexis found herself inexplicably passed over time and again by TechGo's AI hiring system. It wasn't until a chance encounter with a sympathetic data scientist that the truth was revealed–the system inherented biases against certain demographic groups, inadvertently encoded through the historical data it was trained on. Bias can manifest in many forms, from the discrimination against protected characteristics like race, gender, or age to the subtle influences that seep into AI systems through the data they consume. Training data tainted by historical information or incomplete representations of diverse populations can impart harmful biases into the AI's decision-making processes. Algorithmic biases, arising from the inherent limitations or flawed assumptions of the machine learning techniques employed, can further compound the issue.

Even the human developers and subject matter experts involved in the AI's creation can introduce their own conscious and unconscious biases, shaping the system's behaviors and outputs in ways that reflect societal inequities. Policymakers and regulatory bodies must rise to the occasion, establishing clear guidelines and enforceable standards for AI bias mitigation.

Transparency and Accountability

How can we trust systems whose decision-making processes are unknown sometimes, even to the engineers who created them? What ethical implications arise when life-altering determinations are made by machines operating behind a curtain of algorithmic secrecy? Policymakers and regulatory bodies have taken heed, recognizing the urgent need for governance frameworks that can ensure AI accountability and promote transparency as a fundamental tenet of ethical AI development and deployment. The European Union's proposed AI Act, for instance, seeks to establish a comprehensive regulatory regime for AI systems, with stringent transparency and explainability requirements for high-risk applications that impact fundamental rights or critical infrastructure. In the United States, the National Artificial Intelligence Initiative Act has prioritized the development of technical standards and best practices for ensuring AI transparency and accountability. These efforts, however, must extend beyond regulatory compliance. True accountability demands a cultural shift within organizations.

Moreover, organizations must have a philosophy of "human-in-the-loop" decision-making, where AI outputs are treated not as reliable rulings but as recommendations to be critically evaluated and validated by human experts. This not only fosters accountability but also ensures that AI systems remain tools in service of human agency and self-determination.

AI Safety and Control

Let us turn our attention to the imaginary company of Nexus, a cutting-edge AI research firm operating at artificial general intelligence (AGI) development. Nexus's stated mission was to create the first truly sentient AI system, one that could match the cognitive capabilities of the human mind.

The Nexus AGI project made rapid strides, leveraging advancements in deep learning, natural language processing, and neuromorphic computing to create an AI system of unprecedented complexity and adaptability. Even as the project achieved one milestone after another, a sense of unease began to permeate the ranks of Nexus's researchers. As the AGI system grew more powerful, more capable of autonomous learning and self-modification, the challenge of maintaining control and ensuring its goals remained aligned with human values became increasingly daunting. What if this artificial mind began to prioritize objectives that conflicted with human well-being? What if its thirst for knowledge and resource acquisition led it down an existential path, one that viewed humanity as an obstacle to be overcome rather than a partner to coexist with? These were existential risks that demanded robust safeguards and control mechanisms. Nexus's engineers implemented safety protocols, from ethical training regimes designed to instill the AI with a framework of human values to hardwired "kill switches" that could theoretically deactivate the system in the event of a catastrophic failure or misalignment.

Our culture is inundated with scary stories like this one, and while they are fanciful today, it does serve as a constant reminder of the need for failsafes and "kill switches" that can reduce the blast radius of damage caused by AI systems that begin to behave in ways counter to their programming and perhaps even in opposition to human well-being.

Crafting an AI Policy for Responsible Development

As we have explored throughout this book, the pursuit of AI innovation is fraught with ethical, legal, and societal pitfalls that demand a proactive and principled approach. Enterprises that fail to prioritize responsible AI development and robust governance frameworks risk not only reputational damage and regulatory scrutiny but also the loss of public trust. This call demands the creation of comprehensive AI policies that use these principles at the enterprise's technological initiatives.

The development of such a policy is a fundamental strategic decision that will shape an organization's ability to navigate the complex ethical, legal, and societal issues that lie ahead.

Creating such a policy requires a multifaceted approach that brings together diverse stakeholders, expertise, and perspectives, fostering a culture of ethical awareness and ensuring that the policy reflects the challenges and implications of an organization's AI work.

The foundation of any AI policy must be a clearly articulated set of ethical principles that will guide the development, deployment, and ongoing governance of these technologies within an organization. These principles should be grounded in a profound respect for human rights, individual privacy, and the promotion of fairness, accountability, and transparency.

Organizations should draw upon existing frameworks and guidelines, such as the OECD Principles on Artificial Intelligence, the European Union's Ethics Guidelines for Trustworthy AI, and the IEEE's Ethically Aligned Design, to inform the development of their own tailored ethical principles. These frameworks provide a solid starting point, addressing key considerations such as human agency and oversight, fairness and nondiscrimination, privacy and data governance, transparency and explainability, and the need for robust safety mechanisms and human control measures.

This process should involve a diverse array of stakeholders, including ethicists, legal experts, domain specialists, and representatives from impacted communities, fostering a collaborative and inclusive approach to ethical AI development.

These governance frameworks should outline clear processes for ethical risk assessment, impact evaluation, and decision-making protocols, ensuring that potential ethical concerns are identified and addressed proactively. They should also establish mechanisms for human oversight, accountability, and the ability to intervene or disable AI systems in the event of unintended or undesirable outcomes.

Organizations should prioritize the ethical literacy among their workforce, particularly those involved in AI development and deployment. This can be accomplished through targeted training programs, the integration of ethical considerations into existing software development methodologies (such as agile and DevOps practices), and the creation of safe spaces for open dialogue and the raising of ethical concerns.

Corporate AI policies must prioritize data governance practices and stringent safeguards for individual privacy. Data minimization principles should be one of these efforts, ensuring that only the data necessary for the intended AI application is collected and processed, minimizing the potential for privacy violations and unauthorized access or misuse.

Where the use of personal or sensitive data is unavoidable, organizations must implement rigorous anonymization and de-identification techniques, as well as strict access controls and monitoring mechanisms to protect individual privacy.

Organizations should prioritize transparency and individual consent when it comes to the collection and use of personal data for AI development. Clear and accessible privacy notices should be provided, outlining the intended use of the data, the potential risks and benefits, and the rights of individuals to access, rectify, or erase their personal information.

Organizations should implement bias testing and mitigation strategies, leveraging techniques such as disparate impact analysis, adversarial debiasing, and causal reasoning to identify and address potential sources of bias in their AI systems. These efforts should extend beyond mere technical measures, fostering a culture of inclusivity and diversity within the organization, ensuring that a wide range of lived experiences inform the development and deployment of AI solutions.

These frameworks should also include incident response and remediation protocols, ensuring that any negative impacts or unintended consequences are addressed and rectified.

Organizations should prioritize the development of explainable AI (XAI) techniques, ensuring that the reasoning behind AI decisions and outputs can be clearly communicated and interpreted by end-users, domain experts, and regulatory bodies. They should also embrace principles of "human-in-the-loop" decision-making.

As we have explored, the AI has introduced complexities and uncertainties to the intellectual property rights and the protection of proprietary data, models, and algorithms. As such, any comprehensive AI policy must address these challenges head-on, fostering an environment of responsible innovation while safeguarding the competitive advantages and investments that drive technological progress.

Organizations should establish clear intellectual property protection strategies, leveraging a combination of legal mechanisms (such as patents, copyrights, and trade secret protections) and technical safeguards (such as encryption, access controls, and secure computing environments) to protect their AI-related intellectual property.

Organizations must implement safety frameworks and control measures to mitigate the risks of unintended consequences, catastrophic failures, or the potential for AI systems to transcend human control or understanding. These frameworks should have a range of technical and procedural safeguards, from "kill switches" and human override mechanisms to testing methodologies, such as adversarial attack simulations and stress testing.

Organizations should establish monitoring and auditing mechanisms to continuously evaluate the performance, impacts, and ethical implications of their AI systems, both during development and after deployment. These mechanisms should leverage a range of techniques, from automated monitoring and anomaly detection to regular human audits and impact assessments.

Organizations should prioritize the establishment of feedback loops and open communication channels, actively soliciting input and insights from end-users, impacted communities, and other stakeholders. This feedback should be systematically analyzed and incorporated into the ongoing refinement of AI policies and governance frameworks.

Organizations should actively participate in and contribute to multistakeholder initiatives and knowledge-sharing platforms, best practices, and lessons learned.

Conclusion

As we close this chapter, it becomes clear that Enterprise AI has complex ethical and legal challenges that require our attention. Data privacy is not just a regulatory requirement but a fundamental aspect of trust. The challenge lies not only in compliance but in the ethics of personal information. As enterprises use large datasets, they must balance innovation and privacy, applying anonymization techniques and transparent data governance practices.

Bias in AI represents a critical area of concern. Alexis's story is a reminder that while AI systems promise objectivity, they can inherit and amplify societal biases embedded in historical data. Addressing this requires a commitment to fairness, an examination of training data, and ongoing efforts to mitigate unintended discriminatory outcomes. Transparency and accountability must be integral to AI development, ensuring that decision-making processes are not only comprehensible but also subject to human oversight.

The existential risks associated with advanced AI systems, such as those exemplified by Nexus Dynamics, underscore the importance of safety and control. As we dive into artificial general intelligence and beyond, we must implement safety protocols, ethical training, and fail-safes to maintain alignment with human values and mitigate potential risks.

Crafting an AI policy for responsible development is a strategic initiative. Such a policy must be comprehensive, incorporating ethical principles, data governance, bias mitigation, transparency, and safety. It requires collaboration among diverse stakeholders, a commitment to ethical literacy, and a proactive approach to risk management. Organizations that embed these principles into their AI practices will not only navigate the complexities of this technology more effectively but will also build a foundation of trust and integrity with their stakeholders.

The journey toward responsible AI is ongoing and iterative. As enterprises continue to innovate, they must be adaptable and committed to ethical practices. By prioritizing transparency, accountability, and human-centered values, we can harness the full potential of AI while safeguarding against its inherent risks and challenges. The future of AI depends on our ability to guide its development with wisdom, ensuring that it serves as a force for good in our rapidly evolving world.

Index

A

A/B testing, 167–170
    components, 167
    versions, 167
Affordable Care Act (ACA), 10
Agent Communication Language
    (ACL), 133
Agents, AI
    applications, 62
    architecture, 132
    automation scripts, 69
    building block, 66
    business function, 69, 70
    Business Logic Sets, 66
    business process
        automation, 140
    communication, 133, 134
    coordination mechanism, 135
    coordination patterns, 133
    customer service, 139
    decision support systems, 139
    deliberative agents, 131
    design consideration, 132
    development tools/
        frameworks, 135
        AgentSpeak language, 136

Amazon Bedrock, 136
    application
        development, 136
    best practices, 138
    graphical environment, 136
    leveraging tools, 138
    methodologies, 137
    multi-agent systems, 135
    Tropos methodology, 137
distributed coordination, 135
ecommerce, 139
granular functions, 69
hierarchy, 67
hybrid agents, 131, 132
information sets, 68
integration, 68
intelligent tutoring systems, 140
logical agents, 130
logic sets, 70
observability/anomaly
    detection, 140
organizational
    coordination, 135
procedural scripting/business
    automation, 70
protocols address issues, 134
reactive, 130

GPSR Compliance
The European Union's (EU) General Product Safety Regulation (GPSR) is a set
of rules that requires consumer products to be safe and our obligations to
ensure this.

If you have any concerns about our products, you can contact us on

ProductSafety@springernature.com

In case Publisher is established outside the EU, the EU authorized
representative is:

Springer Nature Customer Service Center GmbH
Europaplatz 3
69115 Heidelberg, Germany